SpringerBriefs present concise summaries of cutting-edge research and practical applications across a wide spectrum of fields. Featuring compact volumes of 50 to 125 pages, the series covers a range of content from professional to academic.

Typical publications can be:

- A timely report of state-of-the art methods
- An introduction to or a manual for the application of mathematical or computer techniques
- A bridge between new research results, as published in journal articles
- A snapshot of a hot or emerging topic
- An in-depth case study
- A presentation of core concepts that students must understand in order to make independent contributions

SpringerBriefs are characterized by fast, global electronic dissemination, standard publishing contracts, standardized manuscript preparation and formatting guidelines, and expedited production schedules.

On the one hand, **SpringerBriefs in Applied Sciences and Technology** are devoted to the publication of fundamentals and applications within the different classical engineering disciplines as well as in interdisciplinary fields that recently emerged between these areas. On the other hand, as the boundary separating fundamental research and applied technology is more and more dissolving, this series is particularly open to trans-disciplinary topics between fundamental science and engineering.

Indexed by EI-Compendex, SCOPUS and Springerlink.

More information about this series at http://www.springer.com/series/8884

Shuichi Fukuda

World 2.0

From Working for Others to Working for Yourself

 Springer

Shuichi Fukuda
System Design and Management
Research Institute
Keio University
Yokohama, Japan

ISSN 2191-530X   ISSN 2191-5318   (electronic)
SpringerBriefs in Applied Sciences and Technology
ISBN 978-3-030-51587-4   ISBN 978-3-030-51588-1   (eBook)
https://doi.org/10.1007/978-3-030-51588-1

This Springer imprint is published by the registered company Springer Nature Switzerland AG
The registered company address is: Gewerbestrasse 11, 6330 Cham, Switzerland

Preface

A word to the reader is
"We worked for industry. Now, it is time industry works for us"
This is the main core message of this book.

Current engineering framework since the Industrial Revolution is now approaching its ceiling. And many issues are emerging, such as decreasing workforce, aging population, decreasing childbirth, and decreasing energy. Then, we should explore the new engineering, which will solve these issues and more than that, which enables us to enjoy our own life.

The following are main ideas described in this book:

(1) Human life and machine life are separated. But if we notice that humans and machines live the same way. Both learn to grow and work, then, age. In machines, learning to grow is called break in or fit. And aging is called deterioration. But whatever we may call them, their life is basically the same.

(2) Changes yesterday were smooth so that we could differentiate them mathematically and predict the future. Thus, we could focus on products and paid our efforts to update the functions of a final product. Control was the keyword. Our main efforts were paid to solve the problem faster and with better cost performance.

But today changes are sharp, so we cannot predict the future. Thus, "Adaptability" is getting wide attention. But to achieve our goal, we should do more than to adapt. We need to win the game against the real world.

(3) As IoT pointed out, we used to work outside of the system, but to cope with the frequent, extensive, and unpredictable changes, we need to be a playing manager and must be on the pitch. We need to work together with machines on the same team.

But as soccer demonstrates, other players (machines and humans) must communicate with the playing manager and must be prepared in advance for the next action, which vary widely from situation to situation. In a word, they must be proactive.

(4) What can work for a communication tool between humans and machines, and between machines and machines? It is movement. Humans and machines share movements.

(5) In traditional engineering, results or outcomes are important. And how to produce them is important. So, the goal is fixed. It is problem solving and tactics. Reproducibility is important.

But in frequently, extensively, and unpredictably changing environments and situations, problem finding is important. Or to express it another way, motivation is important.

So, all's well that ends well. We do not care too much how we get to the goal. Reaching the goal is most important. In short, "Pragmatism" will be the philosophy for the next generation engineering.

(6) Traditional engineering was product-oriented and the producer was playing the leading role. Product value was most important.

But next engineering will focus primarily on humans. How we can satisfy human needs is the challenge. Challenge is the core and mainspring to all human activities.

Thus, we need to proceed by trial and error. Life is a game.

(7) Self-Actualization is the highest human need and growth is another important human need. If we pay more attention to "Self", we may be able to expand the new horizon further. And what is important is "Self" engineering not only satisfies needs of individual persons, but it also satisfies the need for human species to expand their world.

(8) To explore such a world, we need to tackle with the problem of computational complexity. But if we observe the octopuses, they die soon after their babies are born. So, they do not inherit knowledge from the previous generation. They live on their own instinct alone. But they can negotiate any environments and situations, and they escape. They are known as "Expert of Escape".

But did we consider instinct in our engineering? I would say "NO". We have been paying attention to brain too much and forgot completely about instinct. But babies learn to move and walk without any textbooks. They use their instinct to learn and grow. If we can use our instinct more, we can tackle with the real world more flexibly and expand our horizons more rapidly and easily.

(9) To achieve such a goal, we need a holistic and quantitative performance indicator to learn how we can improve our performance.

(10) Most of current engineering is based on Euclidean Space approach. It requires orthonormality and units. So, for a small number of dimensions, they work very well, But if dimension increases, it becomes extremely difficult to apply these approaches. Thus, Mahalanobis Distance-Pattern (MDP).

Approach is proposed. Mahalanobis Distance is Non-Euclidean Space approach and it is free from the constraints of orthonormality and units. And it is primarily proposed to identify the outliers. And pattern works for holistic perception.

With the help of this MDP approach, it is expected that we can expand the horizon more easily and establish a new engineering, which provides us with the pleasurable and enjoyable life.

I hope you will read another SpringerBrief of mine, "Self Engineering: Learning from Failures". Together, I believe you will understand my messages clearly. And each chapter of this book is self-inclusive. So, enjoy reading chapters that interest you.

I would like to thank many people for providing me with very different perspectives. I hope we can open the door together to the new world of the next generation engineering.

Finally, I would like to thank Mr. Anthony Doyle and Mr. Werner Hermens, Springer and Mr. Balaganesh Sukumar and Ms. Megana Dinesh, SpringerNature. Their encouragement and patience are truly appreciated.

Tokyo, Japan   Shuichi Fukuda
May 2020

Contents

Chapter 1
Evolving World

We see the figure shown in Fig. 1.1 very often. But usually it is a figure of business cycle called Kondratiev wave. That figure shows how market economy develops [1].

And Joseph Schumpeter is often associated with *Innovation* in market economy and his idea is discussed with Kondratiev Wave, to go up one more step into the next wave. But what he proposed originally is "Neue Kombination", which combines available resources to step forward, as his theory "Creative Destruction" implies [2]. So, he does not mean *Innovation* at all. *Innovation* is the improvement and expansion of the current system. His original idea is not *Innovation*, but *Invention*. He emphasized "Creation".

Although Fig. 1.1 looks just the same as the Kondratiev Wave, our idea is very much different. We are not going to make efforts in the current framework of industry and economy. In short, we do not pursue *Innovation*. We pursue to create a new framework of industry and economy. In short, what we are pursuing here is to invent a new world. So, we share the same idea with Schumpeter.

Here, the world since the Industrial Revolution is defined as "the World 1.0" and everybody is now aware that our traditional industrial world is approaching the ceiling and many issues are emerging.

In this book, we will discuss what the next world, "the World 2.0" will be, and how we can shift from the World 1.0 to the World 2.0. The world before the Industrial Revolution is considered as "the World 0.0" here.

As everybody knows well, before the sixteenth century, i.e., in the age of the World 0.0, *Do It Yourself (DIY)* was too much obvious. If you had a dream and you want to make it come true, you needed to do it yourself. There was no other way. So, you worked for yourself. But situations come to change in the sixteenth century, Work became very much complicated and came to take much time and efforts.

Interestingly enough, the word "work" came to mean "toil" or "forced labor" in the sixteenth century. Until then, work just meant action, which originated from the Greek. So, work until then has nothing to do with forced labor. But since the sixteenth century, work entails the feeling of forced labor. In the eighteenth century,

© The Author(s), under exclusive license to Springer Nature Switzerland AG 2020
S. Fukuda, *World 2.0*, SpringerBriefs in Applied Sciences and Technology,
https://doi.org/10.1007/978-3-030-51588-1_1

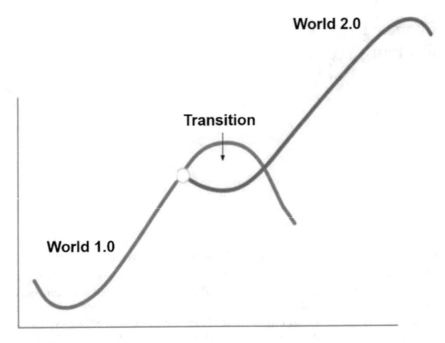

Fig. 1.1 Evolving world

the Industrial Revolution took place, and it introduced the division of labor. Everybody started to work for others, i.e., for external rewards. Most of them did not really know how they were contributing to the design and production of a final product. They just designed and produced as they were told to do.

In other words, we lived until the sixteenth century in the world of *"Self"*. But the Industrial Revolution divided our world, i.e., the world of the producer and the world of the consumer. Such a divided world framework approached its ceiling at the beginning of the twenty-first century.

We must make it clear that "Working for yourself" is not "DIY (Do it yourself)". In fact, the word DIY came from the producer. DIY implies what you can do in the current producer-centric industry framework (the World 1.0).

"Working for yourself" here means to change the framework of industry from producer-centric to customer-centric. In the World 1.0, customers are called consumers and consumer culture [3–6] was generated. The act of consumption became people's major interest in daily life, and they felt satisfied by consuming products. And it brought superiority of social status. And people kept looking for socially higher and better products. Veblen [7], and Riesman [8] also pointed out that in the World 1.0 people pursued such Social Values.

But the World 2.0 is totally different. It really takes care of customers as "customers". In the World 2.0, customers make their own dreams come true in their own "customized" way.

Thus, they are producers, but in a different definition from the traditional industry term. They are not working just for external rewards, but they work based on their internal motivation and self-decision. They work for *Self*. By putting all these diverse *Self*'s together, we can expand our world of human species. In other words, we will evolve as a human. Evolution as species needs diversity. If homogenous, then that species might be destroyed all together at the same time. Diversity is required to expand the world of the species. Growth is an issue of an individual, but evolution is that of the human species.

The World 2.0 is the world in which you make your world happy and satisfactory, and at the same time, you contribute to the world of human species for evolving further and further. We are not sustaining the market economy of the World 1.0. We will challenge for creating another new world in which you can enjoy your life and you help the world of human species expand further. This is a challenge to create a world of psychological satisfaction. The world of material civilization is quickly fading away.

References

1. https://en.wikipedia/.org/wiki/Kondratiev_wave
2. https://en.wikipedia.org/wiki/Creative_destruction
3. C. Lury, *Consumer Culture* (NJ, Rutgers University Press, New Brunswick, 1996)
4. M. Featherstone, *Consumer Culture and Postmodernism* (CA, Sage Publications Ltd, Thousand Oaks, 2007)
5. D. Slater, *Consumer Culture and Modernity* (UK, Polity, Cambridge, 1999)
6. R. Sassatelli, *Consumer Culture: History, Theory and Politics* (CA, Sage Publications Ltd, Thousand Oaks, 2007)
7. T. Veblen, *The Theory of the Leisure Class* (NY, Dover Publications, Mineola, 2012)
8. D. Riesman, *The Lonely Crowd: A Study of the Changing American Character (Yale Nota Bene)* (CT, Yale University Press, New Haven, 2001)

Chapter 2
Our Environments are Changing

2.1 Changes of Yesterday and Today

Today, changes take place frequently and extensively. But we should understand that changes of yesterday and today are completely different.

Yesterday, there were changes, although less frequently and less extensively. But what was very much different was these changes were smooth. So, we could differentiate them mathematically, i.e., we could predict the future. Thus, reproducibility was the keyword. It was important that machines operate the same way, no matter how the situation might change, i.e., robustness was regarded as most important.

Today changes are sharp, so we cannot differentiate them, i.e., we cannot predict the future. Thus, *Adaptability* becomes very important (Fig. 2.1).

Therefore, yesterday, engineers could foresee the operating conditions and design machines that work the same way, no matter how situations might change. Therefore, reproducibility was the keyword and the product functions were important.

Today, however, environments and situations change frequently and extensively, and in an unpredictable manner. So, how we adapt to these changes becomes more important.

In other words, processes become more important than products.

2.2 From Closed World to Open World

Yesterday, changes were smooth, so we could differentiate them mathematically, i.e., we could predict the future. Thus, engineers could foretell the operating conditions, so that they could focus their attention on the functions of their products.

Further, the products were produced on an individual basis, because the world was small and closed with boundary. It was an explicit world so that engineers could apply rational approaches in a straightforward manner. But our world expanded rapidly, and it has changed to an open world without boundary (Fig. 2.2).

S. Fukuda, *World 2.0*, SpringerBriefs in Applied Sciences and Technology,
https://doi.org/10.1007/978-3-030-51588-1_2

Fig. 2.1 Changes of yesterday and today

Yesterday

Smooth Change
Differentiable
Predictable

Today

Sharp Change
Not Differentiable
Not Predictable

Fig. 2.2 From closed world to open world

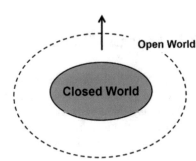

Therefore, we cannot apply rational approaches anymore as we did yesterday.

2.3 Expanding Rational World to Controllable World: System Identification Approach

Engineers solved this problem by developing *System Identification Approach.* The idea of this approach is basically the same as the one we use to identify the name of a river. If we look at the flow, it changes every minute, so we cannot identify its name. But if we look around, we can find mountains, forests, etc. that do not change. Using these feature points, we can identify its name (Fig. 2.3).

By finding such feature points, we can identify the system and we can identify its parameters. Thus, although our world expanded very quickly, we expanded the rational world and established the controllable world (Fig. 2.4).

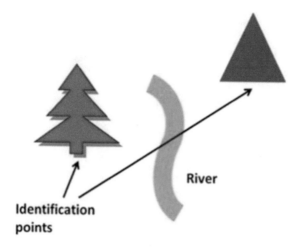

Fig. 2.3 How can we identify the name of a river?

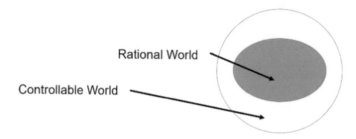

Fig. 2.4 Expanding rational world to controllable world

2.4 Increasing Complexity and Complicatedness

In addition to the problems mentioned above, another problem emerged. That is
Complexity and *Complicatedness*. *Complexity* is very often discussed, but *Complicatedness* is rarely taken up. The discussion about *Complexity* is based on digital
thinking. It is a topic of the discrete world. How discrete elements can be put together
into a system.

But we must remember that when we discuss reliability, the time to break in and
the time to failure or how products deteriorate are very important. These are issues
related to analog world. In fact, nature is composed of digital and analog elements.
Think about our body. Our body is composed of cells, but when talk about our body,
we think of analog parts, such as arms and legs, which are composed of cells. We must
pay more attention to analog phenomena and we should make efforts to integrate
digital and analog worlds.

Chapter 3
Movement is Essential for Living Things

Nobody would object to the idea that *Movement* is the proof that we are living. We, humans, move to live, as the word "Homo Movens" [1] indicates.

The word "Motivation" comes from the Latin "movere", i.e., move. And the word "Emotion" comes from the same Latin. It means to move out. So, we are motivated by the environment and the situation and we move out to the real world to step forward. We need to remember how babies grow. They move out and explore the real world and grow.

But we must remember that *Movement* is also the basic function of machines. Machines are robots in the broad sense. Then, why movement is essential for robots. They are nonliving things. This explains why. Karel Capek [2] coined the word "robota" in his science fiction play "R.U.R.". It means "forced labor" and the word "labor" comes from the Latin, "toil" or "trouble". Thus, robots were expected to reduce our physical labor. So, reduction or rationalization of *Movement* was and is an important challenge in engineering.

And in the age of Internet of Things (IoT), we work together with nonliving machines on the same team. Therefore, movement is important for communication between machines and humans. Machines must understand what is on our mind, when we cannot express our thoughts in words. They must be smart enough to understand our metacognition and what we wish to do. They need to be proactive, as described later in 5: Win the game to survive. We need to move from adapt to outsmart. Otherwise, we cannot win the game against the unpredictable changes. This will be described in detail in 5.3 Soccer as an example.

But we should also remember that movement plays an important role in communication between humans, too. In fact, it is a tacit communication tool. Before symbols or worlds were invented, we communicated through movements. Movement provides us with a clue to what the speaker is feeling inside. We can detect speaker's emotion from face and his or her body movements. Further, we can presume what he or she is intending to do. As will be described in 5.3 Soccer as an example, we can "read" the intention of a person through his or her movements. We perceive how his or her metacognition works. Thus, we can be proactive.

S. Fukuda, *World 2.0*, SpringerBriefs in Applied Sciences and Technology,
https://doi.org/10.1007/978-3-030-51588-1_3

Now, let us consider our motion trajectories. Once we come close to the target, we reduce the number of degrees of freedom to control our trajectories. But at the early stage, the number of degrees of freedom is tremendously large so we have no other way to make a decision on how to move by trial and error.

In other words, we coordinate our many body parts and balance our body. We need team working of our body parts to balance and the formation of this team varies from time to time, depending on the context. And as our body builds are different, movements vary from person to person. Thus, we are now entering the age of coordination. How we can team up our body parts becomes very important.

3.1 Movement of a Robot

But up to now, when we talk about the movement of machines, we always take up movements that can be analyzed rationally. The movement of humanoid robots is not human movement at all. Their movement is rational. They reproduce the rational movements, because engineering up to now has been developed on the basis of the idea of control.

However, the world has changed tremendously, and changes become unpredictable. We are now in the continuously changing flow. System Identification Approach does not work anymore. That is why we do not have swimming robots, Therefore, to develop robots that help us get to our final goal, we need to develop another type of engineering.

3.2 Human Movement

Nikolai Bernstein, human motion control researcher, pointed out that the problem of human movement is its tremendously large degrees of freedom. As our body is composed of a large number of parts, our trajectories vary wide from time to time at the early stage [3], (Photo 3.1).

Bernstein emphasized the importance of coordination in human motion control [4].

To explain it in terms of human body, our muscles harden and move together with our skeleton, when we approach the target. So, our trajectories are fixed. For example, Minayori Kumamoto made it clear that a bi-articular muscles play an important role to make our movement flexible and adaptive [5]. And Yasuharu Koike made it clear that when we recognize that we are close to the target object, our muscles harden, and they move together with our skeleton [6]. So, at the latter stage, we only need to pay attention to the movement of skeleton. Thus, parameters can be easily identified, and the movement can be controlled rationally.

Photo 3.1 Cyclogram of hammering

3.3 Motion and Motor: Two Types of Human Movements

Although coordination is regarded important to understand human movement, strangely enough, human movement is studied from two aspects. One is to observe human movements from outside. This is called *Motion*. The other is to observe from inside. This is called *Motor* (Fig. 3.1)

And these two are not integrated. As Motion Control is very often discussed in engineering (Bernstein's study is one of them), we will discuss Motor Control here.

Five senses are well known. But if we pay attention to coordination and balancing, two more senses are usually added as important.

The sixth sense is Vestibular System [7], (Fig. 3.2).

The Vestibular System is associated with the inner ear and tells us about our head position and how we are moving. It is related to the sense of balance and spatial orientation. Why we can walk in the darkness is thanks to the inner ear. When we talk

Fig. 3.1 Motion and motor

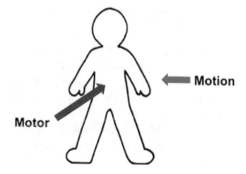

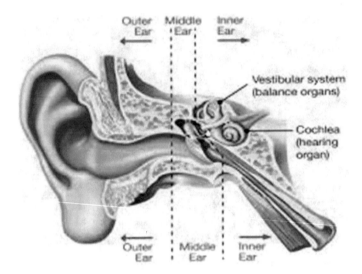

Fig. 3.2 The ear and the vestibular system

about ears, it is about hearing and in most cases, we forget its function of coordination and balancing. It is deeply associated with child development.

The seventh sense is *Proprioception*. It is the sense of the position of the parts of the body, relative to other neighboring parts of the body. It focuses on the body's *cognitive awareness* of movement. And this is associated with *Kinesthesia*. This is awareness of the position and movement of the parts of the body using sensory organs in joints and muscles. Kinesthesia is a key component in muscle memory and hand-eye coordination. It is more *Behavioral* than *Proprioception* [8].

We must remember how we enjoy life when we move our bodies. In an IoT age, we work together with nonliving things (machines) on the same team. So, movement is important for communication between living and nonliving things.

3.4   From Control (Problem Solving) to Coordination (Problem Finding)

The content of this section is discussed later again in Chap. 6 in the broader perspective. The discussion of this section is focused on human movement aspect.

Human movement is very complex and complicated. And what makes the problem very difficult is movement varies from person to person, in addition to the variation of an individual.

Therefore, rational and explicit approaches cannot be applied. In such model-based approaches, the goal is fixed, and we discuss how we can control and get to the goal.

We do our best to find the right road and to drive on a faster lane. It is a problem of tactics.

But in the case of human movements, we have no other choice than to go by trial and error. And we must pay efforts to find the appropriate goal. It is a problem of strategy. Thus, we are now entering the age of coordination. How we can team up our body parts becomes crucially important.

Or in other words, we explore the new frontier every minute in our daily life. When we walk, we perceive the environment and consider the situation, then we make a decision how to coordinate our body parts and move forward. And by trial and error, we find the best way and keep walking.

3.5 From Automate to Harmonize

When changes were predictable and environments and situations did not change much, we could control machines. "Robustness" was the keyword. Machines were expected to work the same way, no matter how the environments and situations changed.

Thus, *Automation* was regarded as an important goal. How we can leave machines to work by itself, without continual monitoring and giving instructions. Once instructions are given, then machines respond the same way without human interaction. In fact, most robots today are pursuing *Automation* and in fact, we desire all machines are automated. Our burdens will be reduced greatly.

The name of "Robot" originally meant "toil" or "forced labor". We wanted to reduce our burdens. That is why robots were invented. But this holds true with all machines. Thus, engineering up to now has been pursuing *Automation*.

Our world, however, becomes more and more complex, complicated and uncertain, so, we need to team up to cope with these situations. And as IoT proposes, we need to work together not only with humans but also with machines. What are needed there is *Harmonization*.

In control-centric world, each member (human, machine) is expected to do best in his own position. In short, society and industry were tree-structured. But today to cope with the unexperienced changes, society and industry are shifting quickly to network. And this network should be truly flexible and adaptable. Members (humans, machines) not only need to play diverse roles, but they need to collaborate together.

Let us consider orchestra. There are many members in the orchestra. But each musician plays his or her instrument. But they need to play together. They need to harmonize. Indeed, you play for yourself, but you play for the whole orchestra, too. This is "work for yourself". It is completely different from "DIY (Do It Yourself)". DIY does not consider how to harmonize with the team. You do it just for you. You think about yourself alone. If you are happy, then that is good. But in orchestra, you need to harmonize with other members, but you need to demonstrate your own capabilities. That is *Harmonization*.

To explain it another way, let us consider dancing. You would like to have a partner who matches you in terms of body builds, etc. You need a static matching first. Then, you need a dynamic matching. You and your partner need to move together beautifully. You need to anticipate how your partner will move and you need to harmonize your movement with your partner.

In an age of IoT, we need smart machines, which move together with us our way, i.e., We need machines that harmonize with us in movement. Then, we feel happy, because we can do what we wish, and we can realize a dynamically matching teamwork.

Automation is the idea of the World 1.0. The World 2.0 is an attempt to create a really harmonizing society. In the World 2.0, we will be happy on our individual basis, but we will also be happy on our society basis. To realize such a society, we also need to pay attention to how we can motivate people (and machines) to realize harmony. Most of our discussions about diversity are from space perspective, but we need to consider diversity along time and space at the same time. In this respect, such new emerging computing as Reservoir Computing is expected to contribute a great deal. It not only serves to attain a high affinity, but it will open door to the new world of sensing, which is indispensable for achieving *Harmonization*.

References

1. K. Kurokawa, *The Philosophy of Symbiosis* (MA, Academy Press, Cambridge, 1994)
2. https://en.wikipedia.org/wiki/Karel_capek
3. https://en.wikipedia.org/wiki/Nikolai_Bernstein
4. N. Bernstein, *The Co-ordination and Regulation of Movements* (Pergamon Press, Oxford, 1967)
5. M. Kumamoto, T. Oshima, T. Yamamoto, Control properties induced by the existence of anatagonistic Pairs of bi-articular muscles–mechanical engineering model analyses. Hum. Mov. Sci. **13**, 611–634
6. Y. Koike, Mapping ECoG channel contributions to trajectory and muscle activity prediction in human sensorimotor cortex. Sci. Rep. **7**, 45486. Tokyo Institute of Technology
7. https://courses.lumenlearning.com/wmopen-psychology/chapter/reading-the-vestibular-sense/ Introduction-to-Psychology,module-5-sensation-and-perception
8. https://en.wikipedia.org/wiki/Proprioception

Chapter 4
Mind–Body–Brain

There are many researchers on this topic. Antonio Demasio, for example, proposed Somatic Marker Hypothesis. Somatic Markers are feelings in the body, and they are associated with emotions [1]. He emphasized the importance of emotion and decision-making.

We often talk about Mind–Body–Brain. There is a word "Make up your mind". But we do not say "Make up your brain". Then, how are they related?

And we should remember that even when our brain is sentenced to death, blood still keeps on circulating out body for several hours. So, within several hours after brain death, we can transplant our body parts to someone else. Brain Science is a hot topic now, but we should remember that mind–body–brain are intricately intertwined

4.1 Mind

When we talk about Mind–Body–Brain, most people think Mind = Brain, without thinking about Body. Very few people really think about Body. But Brain is part of our Body. Why we neglect Body is because when we say Brain, we associate it with Knowledge (Fig. 4.1).

Mind Problem is commonly referred to as Mind–Body Problem. But most people think Mind = Brain and very few people really think about Body. Indeed, Brain is part of our Body, but when we say Brain, it is closely associated with Knowledge. Mind is associated with perception (sensing) and decision-making. Body is associated with sensing and actuation.

Movement is essential for machines and humans. Movements of machines are their functions, and humans are born to move, as we easily understand if we observe babies. We, humans, explore a new world by moving around.

© The Author(s), under exclusive license to Springer Nature Switzerland AG 2020
S. Fukuda, *World 2.0*, SpringerBriefs in Applied Sciences and Technology,
https://doi.org/10.1007/978-3-030-51588-1_4

Fig. 4.1 Mind–body–brain

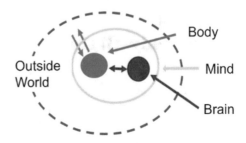

4.2 Direct Interaction with the Outside World

To cope with the frequently and widely changing outside world, direct interaction is indispensable. But the brain-centered traditional engineering has depended too much on knowledge. Knowledge is the structured pieces of our past experience. When situations did not change appreciably, past experience was very valuable for finding a solution. But in a world where unpredictable changes occur frequently and extensively, past experience does not work anymore. What we need now is to perceive the current situation adequately and make the right decision on how to cope with it.

Betty Edwards, American sketch artist, published "Drawing on the Right Side of the Brain" [2]. She tells us that children under 7 years old draw a sketch of a car as they see it. Often they draw a sketch of a car from front. So, you cannot tell it is a car front or a human face. But children after 7 years old draw a car from side. So, you can tell easily it is a car. They subconsciously utilize the fact that our eyes move sideways.

And Antoine de Saint Exupery also describes in his "The Little Prince" [3] that although the Little Prince drew a sketch of a snake swallowing an elephant, all adults who saw, said "It is a hat".

Further, there is an interesting psychological experiment. Hold a group meeting in a room. Then ask attendees to get out of the room. Once all get out, ask them if there is a clock in the room. Half of them do not remember. The other half say, "Yes. There is". Then ask them what type (shape, clock face, etc.) it is. Surprisingly enough, only one or two can answer exactly. All others did not see the clock itself. They saw something that tells time. They saw its function or "concept".

As Edwards points out that after 7 years old, we are not seeing the real world. We are seeing "concepts". So, now you realize that knowledge, which is an accumulation of our past experience is not the accumulation of the data of the real world. It is not objective at all. It is very much subjective. Thus, knowledge varies widely from person to person, depending on his or her way of perception and thinking.

Let us discuss from another perspective of evolution. Interestingly enough, verte-brates and invertebrates evolved in the opposite direction. Vertebrates depend on brains, so they have backbones to support their large brains. Invertebrates, on the other hand, have small or no brains at all. They live on their bodies. The octopus is

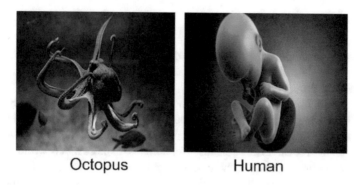

Octopus   Human

Fig. 4.2 Octopus and human

a typical example. Figure 4.2 shows octopus and human and we understand at once how large brains human has and how developed body the octopus has.

The octopus parents die soon after their baby is born. So, there is no transfer of knowledge from parents to children. As past experience is accumulated and structured into knowledge, they have no knowledge at all. They have no other choice than to live on their own instinct.

The octopus, however, is known as the expert of escape. They can escape from any environments and situations. Even when we lock them in a screwed container, they can escape. Humans would be panicked and would not be able to escape. How much the octopus is master of adaptability is described in detail by Peter Godfrey Smith [4]. In other words, the octopus knows very well how to use their bodies. Smith calls it the other Mind. We, humans, on the other hand, rely too much on brains. So, we do not pay adequate attention to our bodies.

Let us return to the discussion about humans. Environments and situations come to change frequently and extensively. And these changes become unpredictable. We are, so to speak, thrown into the flow. We need to find the personal way to swim against the flow and get to our destination, depending on our individual capabilities. Thus, there is no other way than to learn by trial and error. But, how can we?

Engineering up to now is control based, because changes were smooth so that we could predict the future. But now tomorrow is another day. No matter what environment and situation come up, we need to win the game.

If we look at invertebrates, the octopus is a typical example, they can survive on their own instinct alone, without any inheritance of knowledge from the previous generation. But the octopus is known as the expert of escape. They can escape from any environment and situation. This is because they interact with the real world directly.

Our engineering has been too much knowledge centered, but to cope with the unpredictable changes, we need to be more wisdom focused. To achieve this goal, we should pay attention to our instinct and consider how we can utilize it more in engineering to win this challenging game.

4.3 Body

In the case of machines, they operate as designed, but in the case of humans, we need to coordinate and balance our body parts. And we need to do them on our own perception and decision.

Mind is associated with perception (sensing) and decision-making. Body is associated with sensing and actuation. Movement is essential for machines and humans. Movements of machines are their functions, and humans are born to move, as we easily understand if we observe babies. We, humans, explore a new world and learn to grow by moving around. In the case f machines, they operate as designed, and their trajectories are controlled, but in the case of humans, we need to coordinate and balance our body parts to move around. Based on perceived context, we make decisions how to move forward.

4.4 Mind–Brain

Brainstorming is very popular, but there is *Mind storming* as well. But mind storming is not well known as brainstorming. In brainstorming, we produce many new ideas. Then, what do we do with mind storming?

Brainstorming is idea generation, while mind storming is goal finding. In other words, brainstorming is carried out to produce new tactics. Mind storming, on the other hand is to find out goals. We do it to find out the best strategy. This would give you some idea how mind and brain are different.

References

1. https://en.wikipedia.org/wiki/Somatic_markerhypothesis
2. B. Edwards, *Drawing on the Right Side of the Brain* (NY, Tarcher, New York, 1989)
3. A. de Saint Exupery, The Little Prince (New York, NY, Mariner Books, 2000)
4. P. Godfrey-Smith, *Other Minds: The Octopus, the Sea, and the Deep Origins of Consciousness* (NY, Farrar, Straus and Giroux, New York, 2016)

Chapter 5
Win the Game to Survive: From Adapt to Outsmart

As environments and situations come to change frequently and extensively and furthermore unpredictably, *Adaptability* becomes a keyword today. And decreasing birthrate and aging population cause many issues such as decreasing workforce. What makes matters worse, our resources are quickly running out. Thus, *Sustainability* becomes another keyword. But "Sustain" means to keep the present level and "Adapt" means to follow the changes in the outside world.

Changing the topic, growth is human needs. It is individual human need, but we, humans, have another need. Humans need to evolve. It is "growth" of human species. All species of living things survive by evolving. And not only humans, but all living things grow as individuals. Thus, growth and evolution are indispensable for living.

Therefore, *Adaptability* and *Sustainability* cannot deal with the emerging problems. We need to "Grow" and "Evolve". We should pursue *"Outsmart"* and *"Survive"*. We must win the game to survive and to achieve it, we need to outsmart the outside world.

In other words, we need to break away from the current framework of industry and explore the new world. We need to shift from the World 1.0 to the World 2.0.

5.1 From Individual to Team

As contexts become complex and complicated, we need to team up to work out a solution. Two heads are better than one. And not only humans, but machines need to team up to cope with these environments and situations. Further, not only humans and machines need to work as human team and machine team separately, but they need to work together on the same team as IoT indicates.

But in our traditional systems, humans gave instructions from outside of the system and machines responded. Our world and machine world are separated. Such a framework was effective, because the changes were smooth so that we could

S. Fukuda, *World 2.0*, SpringerBriefs in Applied Sciences and Technology,
https://doi.org/10.1007/978-3-030-51588-1_5

foresee the situations. The future was predictable. But as changes became sharp and unpredictable, we, humans, need to be in the system to be fully situational aware.

Let me explain this by taking soccer as an example.

5.2 From Static Team to Dynamic Team

Thus, teamworking is rapidly increasing its importance. Again, we need to integrate human teams and machine teams and let them work together on the same team. Thus, communication between humans and machines become important and movements are very helpful for communication between humans and machines

But we should remember that the current discussion of teamworking is no more effective. Teamworking itself is changing rapidly. Yesterday, there were also changes. But these changes were smooth so that we could predict the future. Therefore, when we talk about team organization and management, we could consider the problem with the fixed number of members and the existing rules. We searched the best solution within these constraints.

In other words, our goal was clear so that we could assume a tree structure and we only had to assign appropriate member to the position. Then we could organize the team.

But today, changes take place frequently and extensively and they occur in an unpredictable manner. So, we need to organize a team that can really cope flexibly and adaptively with these complex, complicated and uncertain environments and situations. Thus, we need to shift from a tree structure to a network (Fig. 5.1). As this network must be truly adaptive, the number of nodes (members) and the number and kind of nodes (how they work together) come to vary widely from case to case.

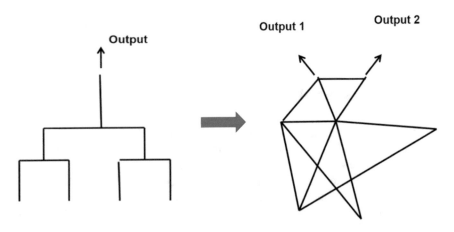

Fig. 5.1 Tree to network

Now we need to organize a truly adaptable team, with substantially no constraints. But we must note that to adapt to the changes that change unpredictably, we need to be prepared in advance for the unpredictable change. In short, we need to communicate with the outside world and anticipate as much as we can what will come next.

In short, dynamic teamworking is required and we need to organize a team dynamically, which can deal with such changes. And further, we must outsmart the outside world. Otherwise, we cannot win the game.

5.3   Soccer as an Example

Soccer is a team sport. So, the formation, or team organization and management, plays an important role. But yesterday, games did not change much, so the basic formation did not change from game to game. The formation stayed the same and each player did his best to perform the expected role at his position. Team fighting strength depended on each player's capabilities.

But as games came to change frequently and extensively, we cannot fight with the same formation. We need to change our formation flexibly and adaptively. To achieve such way of fighting, managers need to be on the pitch. Until then, managers were out of the pitch and they gave instructions. But to understand how the game is changing from moment to moment, they need to be on the pitch together with players and give instructions to them. Thus, today midfielders became playing managers (Fig. 5.2).

And the formation becomes, as the name suggests, the problem of how we form it. So, we do not fight with 11 players as we used to. Sometimes, it is three, or four, member team. We form different types of formation to cope with the changing situations. In short, players need to be multi-competent. They need to play what is expected of them, depending upon the changing situations.

What is important for other players is to be prepared in advance to the instruction from the playing manager. They need to be proactive. Thus, communication is needed. In communication, you must understand what is on the speaker's mind before he or she expresses it as words. IoT is Internet of Things and Internet is the means of communication.

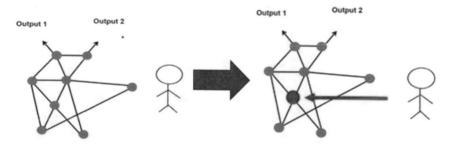

Fig. 5.2 From outside to inside

Knute Rockne, American football player and coach said "11 Best, Best 11". He said "The best team cannot be made by 11 best players. The best team can be made by players who work for the team". He demonstrated this by bringing University of Notre Dame to the ever-winning team. Until he became a coach, University of Notre Dame was at the bottom. But when it moved up to the top, nobody knows the names of players. These players worked together for the team, not for individual merit.

Amy C. Edmondson published "Teaming: How Organizations Learn, Innovate and Compete in the Knowledge Economy" [1]. She emphasized the importance of the dynamics of teaming and tell us to get out of the static teamwork to cope with the dynamically changing context. Although her interest is creating successful enterprises, her idea is basically the same as that of a soccer. Soccer formation has transformed from statically fixed to dynamically changing.

Reference

1. A.C. Edmondson, *Teaminng: How Organizations Learn, Innovate, and Compete in the Knowledge Economy* (CA, Jossey-Bass, San Francisco, 2012)

Chapter 6
From Problem Solving (Tactics) to Problem Finding (Strategy)

Let us discuss from another perspective. Artificial Intelligence (AI) is getting wide attention these days. But we should pay attention to the fact that its cost is tremendously high. Watson, Alpha Go, for example, consume 100–200 kW, whereas human brains consume about 20 W. Thus, AI consumes power 10,000 times as high as a human brain. If we evaluate AI from the point of technological achievements, they are excellent and indeed worthy of praise. But if we look at AI from the standpoint of human satisfaction and happiness, we will realize we need to see things from a much wider perspective.

Tactics is, in short, efforts in each field how to solve the problem faster and better. But strategy is how we can integrate these pieces of knowledge across different fields. It is an activity to find what the problem is and to find the appropriate goal. What is important is transversal perspective.

6.1 Engineering is Changing

6.1.1 Engineering Changes up to Now

Figure 6.1 shows how engineering has changed up to now.

In early days, we focused our attention to individual products, because our world is small, closed and with boundaries. Therefore, engineering was and has been control-centric. And Euclidean Space approach has been used because we can process data mathematically. But our world has been expanding rapidly and there are no boundaries anymore. Now, our world becomes wide, open and without boundaries.

Yesterday, we could apply rational approaches in a straightforward manner. But the expansion of our world brought forth the curse of dimension. Euclidean Space approach has been very effective, because it enables exact mathematical analysis, but high dimensional vector data analysis cannot be carried out because the number of samples is rapidly decreasing due to the rapid progress of diversification and

© The Author(s), under exclusive license to Springer Nature Switzerland AG 2020
S. Fukuda, *World 2.0*, SpringerBriefs in Applied Sciences and Technology,
https://doi.org/10.1007/978-3-030-51588-1_6

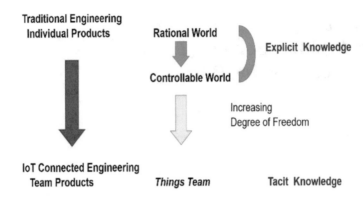

Fig. 6.1 Engineering yesterday and today

personalization. The number of data in the same group is very small. So, more rough approaches are called for. Thus, the needs for a Non-Euclidean Space approach are quickly emerging.

As engineering yesterday was based on the idea of control, automation has been pursued. But automation means that machines operate the same way, no matter how the context or the environments and the situations may change. In other words, robustness was evaluated important yesterday, because changes were small and predictable.

Today, changes are, however, sharp and unpredictable, and they occur frequently and extensively. Therefore, rational approaches cannot be applied straightforwardly. Therefore, we introduced the idea of system identification (See Sect. 2.3). This is the same idea as we identify the name of a river. If we look at the river itself, we cannot identify its name, because the flow is changing every minute. But if we look around and find mountains, forests, etc., which do not change, then we can identify its name. System Identification is based on the same idea. We identify a system, by observing the feature points which follow rational rules. We expanded rational world this way and established controllable world, which is the basis of today's engineering (Fig. 6.2)

But, with the world expanding rapidly, the number of degrees of freedom increases tremendously and in addition, to cope with the diversification and personalization, we are shifting to team product design and manufacturing. Thus, the world is shifting from explicit to tacit and we need multisensory information to cope with this situation. Therefore, coordination and balancing capabilities and decision-making capabilities are increasing importance.

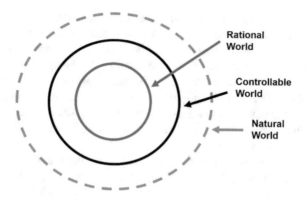

Fig. 6.2 Ration, controllable and natural world

6.1.2   Importance of Instinct

The octopus is a typical example of invertebrates and human is a typical example of vertebrates. They are on the opposite side of evolution tree (Fig. 6.3)

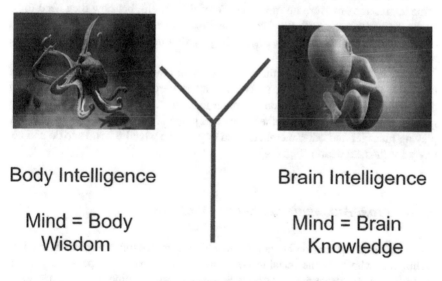

Fig. 6.3 Octopus and human

The octopus teaches us the importance of direct interaction with the outside world. As described earlier, the octopus lives on their instinct alone. They do not inherit any knowledge from their previous generation. So, the octopus interacts with the outside world directly with their bodies. When they "make up their mind", they do that based on their direct interaction with the outside world. Thus, their *Mind* may be interpreted as *Mind = Body* and their intelligence *Body Intelligence*.

Humans, on the other hand, make decisions based on knowledge. So, when they talk about *Mind*, most of them think *Mind = Brain*. So, human intelligence may be called *Brain Intelligence*.

The octopus is excellent in peripheral interactions, but we humans do not have such excellent capabilities. So, *Adaptability* emerged as a keyword. But *Adaptability* is not enough. We need to overcome the natural world, if we wish to make our dreams come true.

In other words, a good strategy is needed to win the game. We have been emphasizing the importance of developing good tactics. But to win, we need to coordinate these tactics appropriately and let them work as a group. Thus, coordination or strategy becomes critical.

6.1.3   Human–Machine Team

Further, machines and humans have been working separately until very recently. But IoT reminds us of the importance of human and machine working together on the same team. Humans were no more outside of the system, but now they need to be in the system to cope with the frequent and extensive changes. And to make such a teamwork better, all members need to be prepared in advance for the next action. Namely, they must be proactive.

We should pay attention to the most important point. Who is going to lead the team and give instructions? In soccer, managers used to do that from outside of the pitch. But today, they are on the pitch to understand the frequently changing game situations. They play together with other players. So, decision comes from this playing manager and other members need to perceive what he wishes to do and get ready for the next action (Fig. 6.4)

6.1.4   From Automation to Smart Machines

When we work together on human and machine team, we must make clear that it is us, humans, who make the decision. And how we work varies from person to person, depending on our preferences, or body builds, etc. So, machines are expected to be smart enough to perceive how we wish them to move and work. Thus, the days of automation are gone.

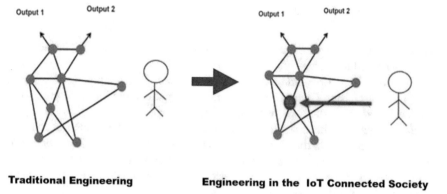

Traditional Engineering   **Engineering in the IoT Connected Society**

Fig. 6.4 From human outside of the system to human in the system

The word *Robots* originally meant forced labor. They were invented to reduce our physical burden. But the etymology of *Machine* is different. It means *Contrivance*. We contrive to make our wishes come true. But unless we do that in our own way, then it is basically the same as working for others. If you would like to work for yourself, then you should expect machines to work in your own way. In short, we come to expect machines to have metacognition capabilities. We want thoughtful and smart machines.

Current robots do not care how we feel or what is your preference of doing the job. They just do the job they are told to do. But machines on the human–machine team need to be thoughtful and provide helping movements to us. These smart machines do not care exactness, as is the case with usual robots. If they can make us happy, that is enough. So, we expect them to act on the basis of human psychology.

We have the word "Make up your mind", but we should be aware that we have subtle *Mind–Body–Brain* relation (Fig. 6.5).

To coordinate and balance such subtle *Mind–Body–Brain*, we need a performance indicator, which can evaluate our performance holistically and quantitatively.

Fig. 6.5 Mind–body–brain

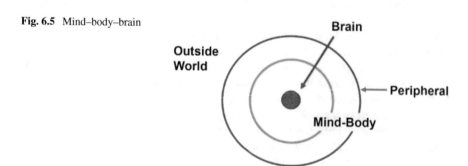

6.1.5 Edge is Important

Here, the importance of *Edge* will be discussed from various aspects, because it is strongly related to strategy.

Recently, *Edge Computing* becomes a hot topic. Just a little while ago, we have been discussing *Cloud Computing*. But we should know that both computing approaches are important, and they have different goals.

Edge Computing becomes important especially from the standpoint of IoT. The term *Edge* comes from the idea of a distributed system, whereas *Cloud computing* represents a centralized system. If we look at the computing resources from the center, they are placed at the edge. So, basically there was no difference between the term *Edge* and the term *Local* in computing.

But IoT added another meaning to *Edge*. Kevin Ashton proposed IoT from his experience from supply chain management. Supply chain is a network. So, Supply chain management is nothing other than the formation and management of an organization. How we can organize and manage network adaptively and flexibly to cope with the changing situations is the top priority issue.

Let us discuss *Edge* from another point of view. *Edge* is deeply associated with object identification, as we easily understand if we think of image processing. Its success depends largely on *Edge* detection. In fact, that is why image processing takes much time and often does not produce satisfactory results. It is very difficult to detect *edges*. If we cannot, we have difficulty in identifying the object.

Edge detection is important in image segmentation. The whole image can be divided into segments by edge detection and we find out what objects there are in the image. So, the image is segmented, and objects are identified.

Well, apart from image processing, what do we do to confirm whether the object is what we think it is or not? We gaze at this unidentified object intently, with the pupils of our eyes fixed and without any movement of them. That is what we observe from outside.

But very interestingly enough. Lawrence Stark, professor of physiological optics at University of California, Berkeley, found out that even in such a case, our eyes are moving hard to detect *Edges*. He is a pioneer in applying engineering control theory to eye movements and he clarified *pupillary light reflex*. He made it clear that even when we gaze at the unidentified object with fixed pupils, the eyes as a whole are working to detect *Edges*. This tells us how important *Edge* is for us.

Speaking of *Edge*, it is the peripheral boundary and we should remember that the octopus directly interact with the outside world with their bodies as described in Sect. 6.1.2. *Edge* in image sensing is related to vision, but *Edge* in the case of the octopus is related to haptics or tactile sensing. As materials are getting softer and softer, we cannot identify what it is by vision alone, we need to grasp it to identify what it is. Thus, haptics is increasing importance these days. In this sense, too, *Edge* is increasing importance. We need to know how *Edges* will change dynamically. *Reservoir Computing* which we will discuss next in connection with *Edge Computing* is, therefore, deeply associated with *Edge* in this sense.

So, although *Edge* and *Local* are used with no distinction in computing, we should be aware that *Edge* and *Local* are very different (Fig. 6.6). The term *Edge* in *Edge Computing* should be interpreted in this sense.

Then, what is *Reservoir Computing*? Why is it gathering attention?

Figure 6.7 shows *Recurrent Neural Network (RNN)* and RNN-based *Reservoir Computing. RNN* is, to put it simply, a directed graph to study temporal dynamic behaviors. Regular neural network is geared to space, whereas RNN is geared to time.

The major feature of *Reservoir Computing* is *Reservoir*. The weights of the links are not variables as in *Neural Network*. In *Reservoir*, they are randomly fixed and at the last stage, weights of the links are varied. This enables large reduction of the number of parameters so that it accelerates online learning capabilities remarkably.

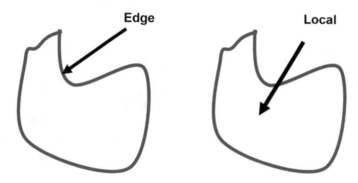

Fig. 6.6 Edge and local

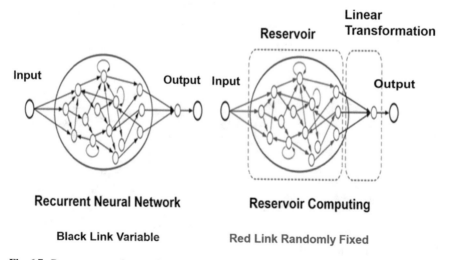

Fig. 6.7 Recurrent neural network and reservoir computing

6.2   Industry Framework is Changing

The changes discussed above is leading us to a new framework of engineering as shown in Table 6.1

6.2.1   From Linear to Network

From another perspective, industrial framework shifted from Linear to Network (Fig. 6.8)

Yesterday, material engineering was not so much advanced as it is today. Therefore, we had no other choice but to select materials among a small number of choices.

Table 6.1 Traditional engineering to strategic engineering

		Traditional Engineering	Strategic engineering
Base		Tactics	Strategy
		Problem solving	Problem definition goal finding
Perspective		Deeper and deeper	Wider and wider
Focus		Time	Time and space
Approach		Model-based	Trial and error
Performance Indicator		Available	Need to find
Value		Function	Adaptability
		Reproducibility	(Evolution)
		Robustness	

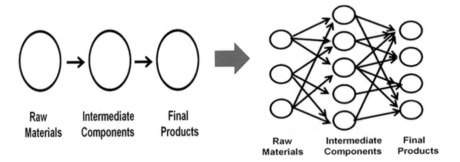

Fig. 6.8   From linear to network

And changes were predictable, so we could set the goal clearly. We wanted products, so we worked linearly toward final products. Therefore, yesterday, final product companies were dominating the market. It was the World 1.0. Everybody worked for others, i.e., for external rewards.

But today, changes occur frequently, extensively and in an unpredictable manner. And material engineering has progressed remarkably. Now, we can ask material engineers what we want and there it is. So, now we have a very broad range of material selection.

A network structure works best for such contexts. And if we introduce *Neural Network*, we can process our work in parallel. Then, we can reduce time and parts companies can keep mass production system, because the same kind of parts can be used for a wide variety of different markets.

Therefore, intermediate component companies are dominating now. But we should remember that each company is focusing on a specific technology. The users combined them differently and come up with different final products.

Thus, from the perspective of workforce, people can do the work that matches their capabilities and preferences. They come to work for themselves. Now, they know what they are designing and producing.

And we do not need many experts. Only limited number of experts who have excellent capabilities in that field is enough. And these experts work for themselves. As Edward Deci and Richard Ryan pointed out in their Self Determination Theory (SDT) [1] that if you are internally motivated and make self-decision, your satisfaction and happiness will be the greatest. No external rewards, i.e., working for others, cannot afford. And Deci and Ryan pointed out the importance of our needs to grow. Experts can enjoy learning and growing in their own fields. So, this is nothing other than the World 2.0.

6.2.2 Increasing Importance of Combination

As described in the previous section, final product companies dominated yesterday, but today intermediate component companies take their place.

When you hear combination of parts, Lego must come to your mind. Lego is a super excellent idea. Although most companies pursue to realize product with better functions, Lego do their business to satisfy customers' human needs. They just design and produce blocks. So, nothing can win in terms of cost performance. And customers enjoy in their own way to combine them into a different final product. In other words, customers enjoy *Self-Actualization* which Maslow proposed and once the product is finished, they are filled with the feeling of achievement. This truly satisfies the human needs SDT pointed out (Fig. 6.9).

Well, indeed, Danish people have rich originality. And they are also famous in design. What interests us is the case of Fritz-Hansen. Their chairs are famous. They produce chairs with excellent artificial leather. But they sell chairs with natural leather

Fig. 6.9 Lego

at higher price. Why? They say, their customers can enjoy the stories about the animals and their life. They sell stories.

Indeed, Danish people have unique culture called Hygge [2]. It is a word for a mood of coziness and comfortable conviviality with feelings of wellness and contentment. They do not worry about the details. They evaluate from a very broad perspective. So, Danish culture may be interpreted as another example of combination. It is fusion of many elements.

Now many industries are introducing the idea of combination. Figure 6.10 shows Daihatsu Copen. Its body panels are interchangeable.

Come to think of it, we change accessories, neckties, etc., when we put on clothes.

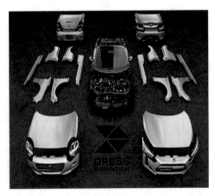

Fig. 6.10 Daihatsu Copen

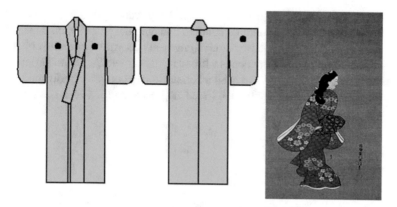

Fig. 6.11 Kimono

Well, speaking of clothes, kimono is nothing but a typical example of combination. It is composed of four fabrics and by re-dying, granddaughter enjoys the kimono her grandmother wore (Fig. 6.11)

And today many "Robots for Kids" toys are coming on the market. Kids enjoy the process of combining parts to produce toy robots.

The idea of combination is old, but it spreads quickly in engineering due to the increasing diversification and personalization.

6.3 Engineering is Different from Field to Field

We should be aware that engineering is very different from field to field. For example, mechanical engineers start from the final product and break it into parts. That is the way they design and produce their products, i.e., top down. But electrical engineers assemble functional parts into a system. So, their engineering is bottom up.

Mechanical engineers like automation and they are always going higher and higher. They do not care constraints. They think all constraints are soft and negotiable. So, if they cannot secure good workers, they will develop robots to replace them. And if humidity is a problem, for example, for welding works, then they introduce air conditioning into the factory.

But civil engineers are completely opposite. They try their best to use local workers. Otherwise, they cannot receive orders from the local community. To employ local people is the best way to finalize orders. And civil engineers know very well things do not go as they plan. In civil construction, it always happens something unexpected will happen. But they have no other choice but to deal with such unexpected events. So, civil engineers are looking at the bottom line. They are always paying attention to what level will be the lowest that the orderer can tolerate.

Civil engineers are truly problem finding. They work with a focus on strategy. But, as environments and situations come to change unexpectedly, engineers in all fields need to work in the same way as civil engineers do. And we need to collaborate across different fields. Strategy needs a broad set of knowledge and experience.

But to achieve such a goal, we need a holistic and quantitative performance indicator. Non-Euclidean Space approach (See Chap. 8) provides a such a performance indicator.

References

1. E.L. Deci, R.M. Ryan, *Intrinsic Motivation and Self-Determination in Human Behavior (Perspectives in Social Psychology)* (NY, Plenum Press, New York, 1985)
2. https://en.wikipedia.org/wiki/Hygge

Chapter 7
Emotion: New Frontier of Movement

Fukuda has been studying Emotional Engineering [1]. His definition of *Emotion* is the following.

Emotion constitutes "Perception ➜ Motivation ➜ Action ➜Emotion" cycle and we move and act by repeating this cycle (Fig. 7.1).

This definition comes from the fact that the etymologies of emotion and motivation are the same Latin word "movere". Thus, emotion means e = ex = out, so, it means to "move out".

Proprioception is described earlier, but to discuss *Motor Control*, *Exteroception* and *Interoception* are important.

Exteroception is the sensitivity of stimuli outside of the body and it is believed the octopus developed *Exteroception* very much. That is why they became the Master of Adaptability.

Interoception is, however, very much related to the human. It is felt experience of the internal workings of the body [2]. *Interoception* is related with emotion. So, if we would like to control motion, we need to consider not only kinematics, but also mental side within ourselves.

Human Movement is not just movement. It is deeply associated with our *Emotion*. Our Brain is more related to logic and reasoning, but *Emotion* is more related to our *Body Movements*. When we say *Body Movements,* most people think about our motion, our body movements observed from outside. But we should remember human movements are composed of two types of movements, i.e., *Motion* and *Motor.*

Motor is our movement inside our body. And we often forgot that blood is deeply associated with *Emotion* and our behavior, as demonstrated by such words as "Blood tingles". And blood carries *Emotion* throughout our body. And we tremble with excitement. And such equipment as Electroencephalogram (EEG) made it clear our brain activities are deeply associated with brain wave patterns.

So, we usually do not observe *Motor Movement* and we still don't know how different parts of our body works together and are coordinated to work best, we should, however, integrate *Motion* and *Motor* movements and should remember that *Emotion* and our movements are deeply associated. So, to detect *Emotion* and to

S. Fukuda, *World 2.0*, SpringerBriefs in Applied Sciences and Technology,
https://doi.org/10.1007/978-3-030-51588-1_7

Fig. 7.1 Perception➔Motivation➔Action➔Emotion cycle

understand *Emotion*, it is indispensable to study our movements. In other words, movement is a tool for tacit communication.

7.1   Self is Our Mainspring

Abraham Maslow proposed the Hierarchy of Human Needs [3], (Fig. 7.2). At the lower level, humans want material satisfaction. But as they go up, their needs change from material to mental. And at the top, they would like to actualize themselves.

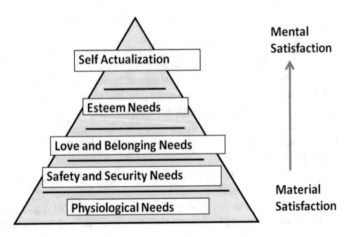

Fig. 7.2 Hierarchy of human needs

We would like to demonstrate how capable we are (See Sect. 7.3 From Rational (Objective) to Self (Subjective): Learning from Failures). This is quite natural. All living things make efforts to expand the world of their species. That is Evolution. To evolve and to expand the world of its species, all living things in that species must have different individuality. If their behaviors are diverse, the possibility that one of them would work to win the game.

Apart from such biological discussion, Edward Deci and Richard Ryan proposed "Self Determination Theory" [4] and they made clear that if we are internally motivated and make a self-decision, then we will have the greatest satisfaction and happiness. No external rewards can provide such satisfaction and happiness, no matter how great the reward may be. Little attention has been paid to such psychological satisfaction and happiness in the traditional engineering. That is because it has focused on products and processes were regarded just a means to produce products. Engineers up to now have only paid attention to product value and they did not really consider the process value.

It is well accepted that engineering started to make our dreams come true. But we should remember that our dreams are different from person to person. So, in this sense, engineering is originally self-centered. Engineering is our activity to bring out our full potential to make our dreams come true. So, it is nothing other than a challenge and we should remember challenge is the core and mainspring of all human activities.

7.2 Decision-Making is Increasing Importance

AI is attracting wide attention these days. But what current *Deep Learning*-based AI can do is to find an answer to the strictly defined problem using big data. Thus, although current AI can solve the problem, it cannot find or discover the problem. In other words, AI can do only what is taught.

But interestingly enough, economists proposed the new sector of the economy. The Quinary Sector in the following Table 7.1 shows that Decision-Making will be the next-generation industry. This sector is very much different from the previous four sectors. These four sectors are based on producer-centric industry framework.

Table 7.1 Five sectors of the economy

Sector	Activities
5 Quinary sector	Decision-making
4 Quaternary sector	Knowledge and ICT industry
3 Tertiary sector	Service industry
2 Secondary sector	Transforms raw materials into Products-manufacturing, etc.
1 Primary sector	Extracts raw materials from nature—agriculture, fishing, etc.

The Primary Sector, which is composed of agriculture, fishing, hunting, etc., is indeed not engineering, but they advanced their industry by introducing engineering. In this sense, it is product- or material-centric industry and it progresses to higher engineering into the Secondary Sector. The Tertiary Sector is a little bit ambiguous. The word "Service" involves many activities. But Service here means the service provided by the producer. Another definition of Service will be explained in connection with the Quinary Sector. And the Quaternary Sector of Knowledge and ICT is the current framework of industry.

The Quinary Sector proposed by economists is very much eye-opening. They pointed out that the importance of executives in companies is increasing. This idea is not an extension of the current definition of the sectors of economy.

To describe it another way, they have been discussing tactics, but now they are moving toward strategy.

But here, we would like to discuss how the new generation industry can be created. Indeed, decision-making is increasing importance, but we are engineers, so we would like to study how it will change our engineering and industry framework, and how it will develop a new world where another value can be created. That is the World 2.0.

7.3   From Rational (Objective) to Self (Subjective)

Let us look from another perspective. In business sector, a phenomenon called Bandwagon Effect is well known. At the time material needs were prevalent, people wanted the same kind of products others have. Thus, the producer paid efforts to produce their products in mass. In such an age, products of the same kind must have the same level of quality. Thus, rational approaches were accepted because the producer can produce the products that meet the expectations of these people. Reproducibility was important because it assures that the product satisfies the *objective* evaluation.

But today, diversification and personalization are progressing rapidly. People come to insist *Self*. Indeed, this is natural. As Abraham Maslow pointed out in 1943 in his hierarchy of human needs, we, humans, first look for material satisfaction at the lower level, but as we go up, our needs shift toward mental satisfaction. And at the top, we would like to actualize ourselves. You would like to demonstrate how capable you are. This is again natural, because all living things are making efforts to expand the world of their species. Humans are no exception. That is how living things evolved and established their own world.

But we must remember that to expand the world of our own species, we need to outsmart the environments. *Evolution* is a phenomenon of a group, but *Growth* is the word for individual persons. *Adaptability* is getting wide attention these days, but it is more than that. We need to *outsmart* or *outwit* the outside world to win the game. In fact, expanding the world of living species is nothing other than a game. We should win the game.

Let us take mountain climbers for example. They select difficult routes, although there are easier routes to the top. And the more difficult the route is, the more satisfied and the happier they are. They are enjoying the challenges or the processes and that is their goal. Getting to the top is not their goal. How they can demonstrate their capabilities is their goal. This is indeed a world of *Self*. But since there is no established route, they have to climb by trial and error.

7.4 Learning from Failures

As trials and errors are inevitable for humans to move, learning from failures becomes very important. Although its importance is emphasized, we do not know how. This is because we do not have an appropriate performance indicator. The learner cannot understand how he or she is doing better this time.

If a performance indicator is not provided, we have to try and error without any strategy. Then, our trials and errors will be unorganized, and we cannot improve our performance (Fig. 7.3).

But if an appropriate performance indicator is available, the learner can gradually narrow the scatter and keep the scatter within the reasonable range (Fig. 7.4).

So, learning from failures becomes important. "Fail" here means "fail one's expectations". This is none other than *Pragmatism*. But although its importance is emphasized so much, there is no, at least to the author's knowledge, quantitative performance indicator available.

In the following, performance indicator for evaluating movements quantitatively and holistically is proposed.

Fig. 7.3 Unorganized trials and errors

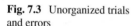
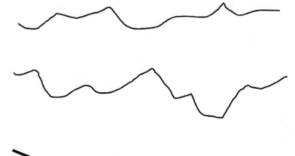

Fig. 7.4 Organized trials and errors

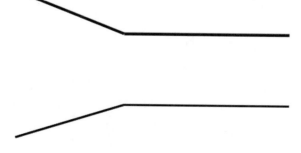

We should make efforts to create such Pragmatic world. That is the World 2.0. The World 1.0 was a consuming society, but the World 2.0 is a self-producing society. The word "Self-producing" might be misleading. You might imagine such an equilibrium state as in autopoiesis. So, let us call it "Self-growing Society". It is a Society that keeps on growing without substantial support from outside. And these individual growth leads us to evolution, and we will expand the world of human species.

References

1. S. Fukuda (ed.), Emotional Engineering, vol. 1–8 (Springer, London, 2011–2020
2. http://en.wikipedia.org/wiki/Interoception
3. A.H. Maslow, A theory of human motivation. Psychol. Rev. **50**(1), 370–396 (1943)
4. R.M. Ryan, E.L. Deci, *Self-Determination Theory: Basic Psychological Needs in Motivation, Development, and Wellness* (NY, The Guilford Press, New York, 2017)

Chapter 8
Performance Indicator

What is lacking in the current industrial framework is a consideration on how they can provide their customers with psychological satisfaction. The current engineering is focused on products. And it regards customers as consumers. The current industrial framework emphasizes consumer culture. But as the world is expanding rapidly and it is now an open world without boundary, diversification and personalization are emerging as a key concept. To cope with this shift, we should move away from mass production and consumption to diversified and personalized customization. We should change our industry framework to meet such personal and mental requirements. And as Deci and Ryan pointed out the maximum satisfaction and happiness are achieved when we are internally motivated, and we make self-decision.

Thus, engineers should make efforts to help their customers find out what satisfies their personal and mental desires and to let them believe they find out the solution by themselves. Engineers are behind the scene and help them. This may be compared to the relation between parents and a baby. A baby plays the leading role. Engineers should think out how they can let their customers feel they are playing the leading role. Now is an age of "*Self*" Society and engineers must convince their customers that now is age of "*Self-Satisfying Society*" (Fig. 8.1).

To achieve this goal, each customer needs to know how he or she is doing better this time. *Growth* is an important motivation and it drives them to make efforts to satisfy themselves. In fact, when they achieve their goal, they will have the feeling of achievement and it motivates them to challenge one step higher goal. This is *Growth*.

Growth may be described as communication with the real world. We learn how to deal with the real world. This is *Growth*. Babies communicate with the real world and learn and grow. Movement may be called tact communication. Babies learn to move before they learn to speak. It is *Instinct*, which engineers have forgotten to pay attention to up to now.

Many equipments are available today for measuring movements, but they show the perfect movement. Our body builds are different from person to person and how we make decisions also vary from person to person. So, we need a performance

S. Fukuda, *World 2.0*, SpringerBriefs in Applied Sciences and Technology,
https://doi.org/10.1007/978-3-030-51588-1_8

Fig. 8.1 Baby exploring the
real world

indicator which works for each one of us in his or her own way. Babies grow in their
own way.

8.1 Euclidean Space Versus Non-euclidean Space

First of all, let us consider how difficult it is to deal with movement of living things.

If we pay attention only to skeletons, we can identify parameters. So, Euclidean
Space approach and Euclidean Distance can be applied in a straightforward manner.

But now, environments and situations change unpredictably. In other words, we
are now thrown into the flow. Water is changing every minute. We cannot identify
any parameters, but we need to swim against the flow to reach our goal. As we
experienced very well in swimming, muscles play an important role. But muscles
are soft and deformable. That is why haptic are increasing importance in grasping.
And our muscles vary widely from person to person. Further, we do not know well
what constraints there are about muscles.

So Euclidean Space and Euclidian Distance do not work anymore in such soft environments. We need Non-Euclidean Space approach, which is free from the requirements of orthonormality and units. Besides, we do not know how different parts
of our body work together as a team to realize such performance as swimming.
Mahalanobis Distance (MD), which will be described in the next section provides a
solution to this challenge.

To describe this in mathematical terms, Euclidean Space calls for normalization.
This is related to the problem of units. And regularization is required to process

high-dimensional vector with few samples. This is the case Occam's razor or law of parsimony comes up. But normalization and regularization are required if you use Euclidean Space approach. If you use Non-Euclidean Space approach, you will be free from these requirements. This is what is described in 8.3 Pattern: Holistic Perception. Instead of processing all data mathematically, it is proposed in this book to use our *Instinct* more for judging the pattern. We see the images and classify the pattern on a holistic basis. Or it may be called the ultimate case of Occam's Razor of the Law of Parsimony.

8.2 Mahalanobis Distance (Md)

P. C. Mahalanobis proposed Mahalanobis Distance (MD) in 1936 [1, 2]. It is a measure of distance between a point P and a distribution D. Although Mahalanobis is a researcher in the field of statistical testing, this idea has nothing to do with statistical testing. His goal was not to test his hypothesis using statistical approach, but to identify if a data is an outlier or not. It is the multidimensional generalization of the idea of how many standard deviations away P is from the mean of D (Fig. 8.2). MD is expressed as

$$MD = (\text{standard deviation})/(\text{mean}) = \sigma/\mu$$

As his primary purpose is to determine if the data is the outlier or not, the datasets do not have to be normally distributed. MD can be applied to any form of dataset distribution and further datasets can be independent of each other. Thus, MD can be applied to multidimensional problems easily and it can increase dimensions without any difficulty and process them in a very short time. This will be explained in detail later in 8.4 Mahalanobis-Taguchi System (MTS).

Fig. 8.2 Mahalanobis distance

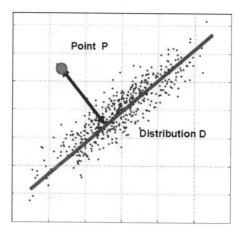

8.3   Pattern: Holistic Perception

Artificial Intelligence (AI) is attracting wide attention these days. Especially *Deep Learning* is a hot topic. But we should remember that the basic idea of *Deep Learning* was developed about 50 years ago.

At that time, AI was called *Expert Systems,* because *Knowledge* itself was taken care of by experts and the role of AI was to provide a platform for implementing expert knowledge. MYCIN was developed by Edward Shortliffe at Stanford and it established the name of expert systems in the early 1970s [3]. This was a very clever strategy, because *Knowledge* varies widely from field to field.

But soon it was taken over by the name of *Knowledge-based Systems* and each system was developed for a particular field of *Knowledge* and *AI* is called *Knowledge Engineering.* And *Deep Learning,* which is based on *Hierarchical Neural Network,* was developed. We should be aware that the basic framework of *Deep Learning* has not changed much since then. Why it came back into the spotlight is not due to the progress of software, but that of hardware. Computer hardware has progressed remarkably, so that what cannot be processed 50 years ago can now be processed easily. Thus, AI today is good at finding a solution using big data to the adequately defined problem. AI 50 years ago learned how to solve the problem using human knowledge. Thus, if humans could not find the solution, AI could not. But today, *Deep Learning*-based AI can find a solution among big data, which humans cannot explore. And pattern recognition plays an important role in AI to find a solution using big data.

But the capabilities of AI are still limited.

(1)   It can only solve the problem given by humans,
(2)   The problem must be clearly defined,
(3)   There are strict rules.

What AI today cannot do are the following.

(1)   To find problems,
(2)   To solve the ill- or poor-defined problems,
(3)   To define a problem using small data, and
(4)   To create new rules to adapt to the environments and situations.

Thus, although problem discovery is increasing importance in this age of complexity, complicatedness and uncertainty, AI cannot contribute at all in this respect. But interestingly enough, although computer systems have been moving toward centralized systems such as *Cloud Computing* until very recently, they are now quickly moving toward distributed systems due to the invention of the idea of *Internet of Things (IoT),* and *Edge Computing* is now attracting wide attention. IoT originated from the needs of supply chain. So, we should remember that it is based on discrete mathematics.

Until now, *Controllability* was the keyword, so continuous mathematics or rational approaches have dominated engineering. But since IoT is teamworking of supply chains or network, it allows ambiguity or vagueness. But we should be careful. *Fuzzy Theory* is different. It is a many-valued logic. Discrete mathematics is not a logic and it can allow rough definition and processing of the problem.

As IoT accelerates *Edge Computing*, because small sensing or actuating devices come to be needed at each *Edge*. But *Neural Network* and *Recurrent Neural Network* take tremendous time and energy. Watson and Alph Go consume 100–200 kW, while human brain consumes approximately 20 W. To cope with this problem, new Recurrent Neural Network called *Reservoir Computing* is proposed [4, 5], which only performs training at the readout stage.

If we look at it from the other way, images can be turned into pixel sequence or time series. This *Reservoir Computing* changes time series data into patterns. So, if we look at patterns (The name *Reservoir* came from the association that when a stone is thrown into the pond, it creates ripple patterns), we can roughly or holistically come up with the training data and train the system. In other words, we adjust to meet our expectations at the last stage only. Thus, large reduction of time and energy can be achieved.

It is our *Instinctive Capabilities* that we can distinguish holistic pattern characteristics.

And *Reservoir Computing* contributes to the development of hardware devices, which is the other way around from traditional computing. *Reservoir Computing* is one of *Natural Computing*, which characterizes nature or the real world.

Although it is a digression from the main theme of this book, we should remember that images play very important roles in medical diagnosis. Such machines as PET, fNIRS, fEMG, ECG, and EEG use images. Medical diagnosis needs holistic judgment and the discussion here is expected to be useful for medical applications, too.

It should be added that understanding and processing patterns holistically is expected to serve a great deal for many applications. And Mahalanobis Distance-Pattern (MDP) approach proposed here will lend a very effective helping hand.

8.3.1 Detection of Emotion

Fukuda used to study detection of emotion from face. His group tried many different image processing techniques, but they were too much complicated and took too much time. And the results were not satisfactory.

During these efforts, he realized that we could recognize emotions of characters in cartoons without any difficulty. At that time, cartoons were in black and white, and characters were very simple. But we could detect their emotions at once. So, Fukuda developed a cartoon face approach and his group succeeded in detecting emotion from face easily in a very short time (Fig. 8.3), [6, 7].

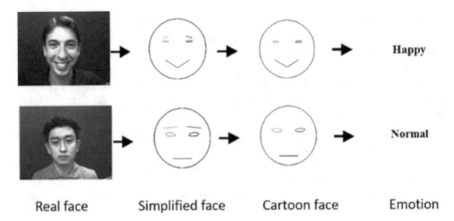

Fig. 8.3 Cartoon face approach

8.3.2 Instinct

As this cartoon face research was a model approach, we used a cluster analysis. So, it was carried out in *Euclidean Space* with *Euclidian Distance*. But if we consider cartoons again, we are not applying any model to detect emotions of characters. No models, but still we can detect emotions. And we have not seen the characters before. But we can detect their emotions. This is nothing other than *Instinct*. We are born with such *Holistic Perception* capabilities. Our traditional engineering has forgotten to pay attention to such *Instinctive* approaches.

8.3.3 Non-destructive Inspection (NDI)

In fact, engineers are aware that such instinctive or holistic tools are missing. For example, in Non-Destructive Inspection (NDI) field, "walkthrough" is being carried out in plants to detect flaws. And in aviation field, pilots and aircraft mechanics walk around the plane before flights. In fact, we share the same experience in our daily life. When we walk into the room and feel something is different, we watch carefully and find flaws on the wall, etc. Such *Holistic Sensing* or *Holistic Perception* has been forgotten or very few efforts have been made up to now.

8.3.4 Medical Diagnosis

Such *Holistic Sensing* or *Holistic Perception* is very important in medical diagnosis, too. There are many general hospitals. And all of them are divided into medical departments. But they cannot carry out *Holistic Sensing* or *Holistic* diagnosis. They are divided, not united. This is exactly the same as in NDI. If a flaw can be detected, then surprisingly accurate analysis can be carried out, but they cannot answer the simple question, "I feel something wrong. What will it be?" This is because there is no, at least to my knowledge, *Holistic Sensor*.

8.3.5 Shoji (Japanese Paper Wall): Cellulose NanoFiber (CNF)

Interestingly enough, Japanese "shoji" or paper wall works as a *Holistic Sensor* (Fig. 8.4). Japanese paper is made by the same idea of CNF (cellulose nanofiber), which is expected to be the material for the next generation after the current CNF (carbon nanofiber). Western way of papermaking is layer based, such as being done in additive manufacturing, one approach of 3D printing, which successively adds layer by later.

Thus, current western-type walls separate the inside and outside of the house. But Japanese "shoji" does not. Even when we are in the home, we can feel the atmosphere of the outside. If the wind is blowing, we can feel the wind. Wind, sound, humidity, etc. come through the shoji. So, when the new CNF comes to be used widely for vehicle bodies, housing walls, etc., we can feel the atmosphere of the outside world. We can feel the outside, even when we are in the car.

Fig. 8.4 Shoji

8.3.6   Granular Material Engineering

It should be added that CNF (cellulose nanofiber) is one typical example of granular material engineering. And in this field, there are a large number of new horizons. Granular material engineering is used in medical, inkjet printer, air conditioning, etc. in a wide variety of fields. For example, in air conditioning, instead of airflow, hitting us by very small particles makes us feel cool. So, we can develop another type of air conditioner. And children play on the seashore with sand (Fig. 8.5). We can develop another type of toy instead of Lego type. And it should be emphasized this is truly recyclable.

8.3.7   Jack of All Trades

And in Japanese architecture, there is no distinction between shoji and wall. So, we can change room layouts easily. On the other hand, columns and walls are assigned different roles in western architecture. The columns carry weights, but walls do not. So, it is not easy to remodel the room in western architecture. Japanese culture is holistic and does not pay any particular attention to details. But western culture is self-assertive and each element has its own role.

Fig. 8.5   Child playing with sand on the beach

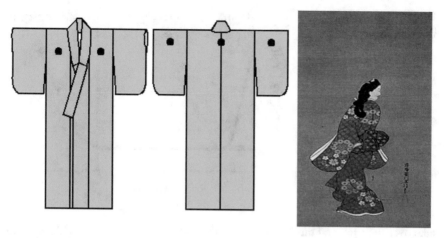

Fig. 8.6 Kimono

8.3.8 One Size Fits All

Such cultural difference can be observed in fashion field, too. Western dresses are made to fit the body of each person, but Japanese dress, Kimono (Fig. 8.6) is just like a bathrobe. Japanese just put on kimono. They do not wear it closely on the body. And kimono is made up of four fabrics so that by re-dying and combining them in a different way, granddaughter can enjoy the same kimono her grandmother wore.

8.3.9 Break In: Fitting Over Time

If customers can identify their needs correctly and convey them to the producer, they can answer to that. In reliability engineering, bathtub curve (Fig. X) is well known. It shows the change of failure rate over time. But we should note that some time is needed for break in and after break in period. Products need to adapt to us, users and it takes time. After break in period, they really start to work for us. *User Experience (UX)* has been discussed very often, but most of the discussion is from the standpoint of the producer. UX Feedback from the user has been discussed, but these ideas did not work well.

Well, the word "work" is used now. It would interest you to know that nearing the end of seventeenth century, the word "work" come to mean "torture", "toil". The Latin etymology has such negative implications. But the Greek origin, on the other hand, does not have any such negative nuance. On the contrary, it sometimes means "joy". Why did "work" change from positive to negative? This may be because work became too much physical labor, so it gradually changed from joy to toil. And at the beginning of the twentieth century, the word "robot" was coined, because work then becomes none other than "forced labor", so everybody would like machines to

Fig. 8.7 Bathtub curve (CMU EME)

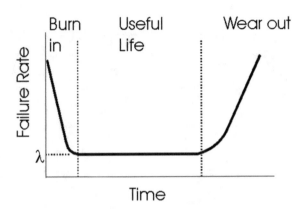

take over their job and the Industrial Revolution introduced *Division of Labor* and everybody started to work for others.

Awareness or perception characterizes humans. AI is devoid of such capabilities.

That is why decision-making is an important role in humans. But if we look at Fig. 8.7, we should be aware that what users want is the performance of the product during useful life.

ASICS, Japanese shoemaking company, realized that all worn out shoes are deformed as shown in the middle of Fig. 8.8. Until they became aware of this, they produced shoes in the same way as other companies, i.e., they produced shoes with the initial performance quality in Fig. 8.7. They did their best to keep shoes strong enough so that the shoes do not deform easily and will not be thrown away soon. But what they really wished was to extend the useful life of their shoes as long as possible (Fig. 8.7).

But on second thought, they realized that in bathtub curve, the initial period is called burn in (we call it break in) and when the product comes to fit, then, the product is put to regular use. So, if the shoes deform that way, that means, that is, the form people wear for their daily use.

ASICS recognized the importance of their discovery. They changed their design and modularized the shoes. They made the middle part easier to deform. These shoes were welcomed so much by customers. They say "I feel like walking on my feet. I do not mind walking a long distance. I enjoy walking".

Fig. 8.8 ASICS modularized shoes

This success reminded ASIC that at a start of running, the same deformation takes place. So, ASICS designed new shoes for runners. These successes were brought forth by *Holistic Perception*. ASICS perceived holistically how their shoes are getting customized. They identified the characteristics of human foot movements as a feature pattern.

8.3.10 Movement: Mind Mining

Data Mining is attracting attention, because deep learning-based AI is now quickly spreading its applications. It discovers the pattern from big data. But patterns used this way is pattern recognition. What is discussed here is related to Metacognition. *Holistic Perception* implies to understand the partners' Metacognition. We need to understand what the speaker or the team member has on his or her mind without words. If it can be expressed as words or symbols, then it boils down to *Data Mining*. *Data Mining* is based on Euclidean Space. So, such items as normalization, regularization, cluster analysis, etc. come up. But in the case of *Holistic Perception*, fundamentally we rely on *Instinct* for decision-making, so we can avoid the curse of dimensionality. In fact, it is not only the problem of dimensionality, but it is the problem of time and dimensionality. So, in other words, we can purse the Law of Parsimony if we combine MD with pattern. And if we pay attention to materials, they are getting softer and softer. In short, hardware is shifting to software. So, our world is now moving toward soft.

Therefore, *Mind Mining* is emerging. We need to anticipate the action or behavior of the speaker or the team member. We need to understand what he or she is perceiving and what decision he or she will make and what action he or she will take. Then, we can establish real communication, not only between humans, but also between humans and machines. Movement reveals the metacognition of the speaker or the team member. Movement plays a very important role in *Natural Computing*.

8.3.11 Matrix is a Pattern

Thus, patterns are important from the standpoint of *Holistic Perception* or *Holistic Sensing/Feeling*.

When Fukuda came across Taguchi's pattern approach using MD (MTS), he realized that if he introduces MD, then he does not need to use cluster analysis. He can develop a holistic performance indicator without worrying about the constraints of Euclidean Space.

It should also be added that a matrix is nothing other than a pattern. As described in the next section of MTS, a matrix can be represented as a pixel sequence pattern. So, we can process matrix not in a quantitative way as we usually do, but as a pattern to grasp the holistic image. In other words, matrices can work both ways for quantitative and holistic evaluation. In fact, the word *Pattern* comes from the Latin "model" and *Image* comes from the Middle English "imitate". Thus, matrices are very useful for identifying the model not through System Identification Approach, but through *Intuitive Model Identification.*

It should be pointed out that color images can be represented as matrices. And color matrix can serve as a communication tool beyond words. When we see images, we can share our feelings. These feelings are, so to speak, subconscious so they are difficult to express in words, but we can certainly share the feelings. This is because in such cases we are processing matrices as a pattern, not as a mathematical representation. And as MD is non-Euclidean, it facilitates such flexible processing.

8.3.12   Focused Versus Holistic

Patterns are images. Images provide us with a holistic and synthetic perception, because we look at the whole space. But we should take care of that vision which we talk in connection with the five senses is different. When we see, we focus on the target. We make a decision and look at the target in particular. So, it is very much subjective. But images are signals or stimuli from outside. So, it is objective.

While images are the world of synthesis, letters are that of analysis. Image data are scattered all over space, whereas letter data come up sequentially, so we need to follow them one after another. We need to focus our attention on letters to identify what letter it is. Thus, images are spatially distributed, but letters are temporally distributed.

We must take note that the same phenomenon is observed in hearing. Usually hearing is understood to follow signals one after another. That is the interpretation of hearing as one of the five senses. But as our ears are open all the time, when we notice something is different, it is the same as we experience in images or patterns. We are holistically context-aware in such cases.

8.4   Mahalanobis-Taguchi System (MTS)

Mahalanobis's main interest was to remove outliers, but Genichi Taguchi noticed if the quality elements are represented as a pattern, then MD can be used for pattern identification [8].

Figure 8.9 shows the basic idea of *Mahalanobis-Taguchi System (MTS)*.

Unit space is defined as shown in Fig. 8.10. The ideal pattern is *Unit Space*. Then, he introduced the Threshold. If MD is smaller than the Threshold, then the quality is acceptable. If MD is larger than the Threshold, we need to examine and determine which element should be improved.

It should be noted that a pattern can be generated using pixels. So, images can be analyzed as well (Fig. 8.11).

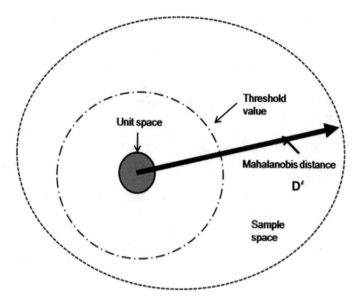

Fig. 8.9 Mahalanobis-Taguchi system

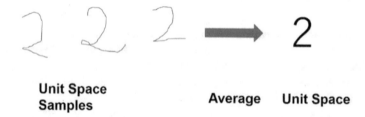

Unit Space Samples   **Average**   **Unit Space**

Fig. 8.10 Definition of unit space

Fig. 8.11 Pixels to patterns

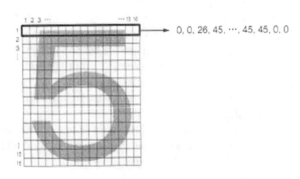

0, 0, 26, 45, ···, 45, 45, 0, 0

MTS is not a statistical test. It measures how far away the point P is from each dataset on an individual dataset basis. In a statistical test, a hypothesis is tested statistically. But Taguchi noticed that if he combines MD with pattern, we can carry out *Holistic Evaluation*. In a hypothesis testing, it is carried out element by element. *MTS* compares patterns and evaluates how close or far away the pattern is from Unit Space. So, evaluators do not have to care about each element. *MTS* is *Holistic Perception* of a pattern.

Thus, *MTS* is welcomed by many companies, because they cannot control quality element by element. They can only monitor and control quality holistically. This is possible, because MD itself is Non-Euclidean, so it is free from the constraints of orthonormality and units.

Genichi Taguchi and Rajesh Jugulum published "The Mahalanobis-Taguchi Strategy: A Pattern Technology System" in 2002 [9]. The title of this book is a little bit confusing. This book is purely based on the idea of statistical testing. So, Taguchi calls this system *Mahalanobis-Taguchi-Schmitt* because it is based on the idea of Euclidean Space. Thus, it requires orthonormality and units, whereas *MTS* does not.

8.5 Pattern: From Static to Dynamic

But human movement is dynamic. And if we consider swimming as an example, we immediately realize we cannot find any feature points as we did in *System Identification*. Water is changing from moment to moment and the style of swimming varies widely from time to time and from person to person. So, each learner has to learn by himself or by herself and by trial and error.

MD is unitless and just defined as (standard deviation)/(mean), so we can apply MD and pattern approach, no matter how widely diversified the swimming style is. The learner recognizes how much he or she is improving or deteriorating. In short, it is a dynamic pattern evaluation.

If the learner starts to swim and there is no successful pattern of his or of her own, then we can use someone's successful pattern as a substitute. And as the learner begins to succeed, we can update the *Unit Space*, so he or she can evaluate his or her own improvement or deterioration in a more detailed manner.

And as images can be represented as a pattern, we can process dynamic images. So, we do not need to put sensors on swimming suits. Of course, we can, but we can do without it.

Swimming is a very good example of tacit dimension and it illustrates how it is important for us to interact with the outside world. And further, our body is soft. So, a *MD and pattern (MDP) Approach* proposed here will be very much useful in the emerging world of soft materials.

One more important point should be added. It is the role of acceleration. Fukuda and his group studied calligraphy and found out that acceleration of the movement of the writing brush exhibits the writer's personality. We usually only think about trajectories. In fact, Bernstein and other motion control researchers discuss only human motion trajectories.

But why we can identify who signed it is because it not only represents brush or pen trajectories, but the characteristic of acceleration can be recognized. In fact, humans are very much self-conscious, but such self-expression of emotions is not recognized enough. So, just the static pattern matching goes for signature in most cases. But if we identify how the brush or the pen is accelerated by looking at the signature, we can certainly identify the person who signed it. This is *Dynamic Pattern Matching*.

In the case of detection of emotion from face, Paul Ekman and Wallace V. Friesen proposed *Facial Action Coding System (FACS)* and pointed out the importance of face muscles [10]. But their research did not go beyond static pattern matching. But we can understand the emotion of the other party better, if we observe how his or her face changes, i.e., we understand his or her emotion in terms of the acceleration of face muscle movement. In this sense, movement is a tacit communication tool, but often it carries messages more than just words.

Mahalanobis Distance-Pattern (MDP) Approach developed and proposed here is a performance indicator with particular attention paid to the acceleration of movements.

8.6 Systems: From Centralized to Distributed

With increasing diversification and personalization, computing systems are rapidly shifting from centralized to distributed. Until very recently, such centralized systems as *Cloud Computing* has attracted wide attention. But we should pay attention to the fact that such a new technology as *Edge Computing* is emerging. In *Edge Computing*, computation is carried out on small devices at the edge. These devices do not consume a large amount of electric power such as Watson or Alpha Go. It would surprise you to know that they consume 100–200 kW. Humans, on the other hand, consume only 20 W.

The needs for *Edge Computing* are growing rapidly, not only because diversification and personalization accelerate localized computing, but because small devices are needed for sensing. For example, devices on the drone cannot consume such large electric power or space. To solve such problems, *Reservoir Computing* is proposed.

In Artificial Intelligence (AI), *Deep Learning* is attracting wide attention these days across many fields. But it is based on *Hierarchical Neutral Network*. In other words, computer searches big data for a solution. It is based on space pattern. The feature of space pattern is identified.

To expand pattern recognition beyond just space and toward over time, *Recurrent Neural Network (RNN)* is proposed [11]. It processes one pattern after another over time and identify the feature of time-space pattern. The basic idea of RNN is illustrated in Fig. 8.12.

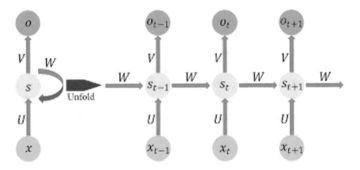

Fig. 8.12 The basic idea of RNN (Researchgate.net)

As our society needs for *Edge Computing* and *RNN* increase, such emerging idea as Internet of Things (IoT) calls for small sensing devices to work locally. It is very natural because IoT was originally proposed to cope with the problems of supply chain. Supply chain cannot be managed effectively by centralized system. It works better to leave management and decision to each local chain. And if small device sensors can be developed, they can make their decision faster and more effectively on site. And needless to say, supply chain management needs temporal pattern features to manage effectively.

Thus, *Edge Computing* and *RNN* go very well with IoT. But *RNN* needs a great amount of computing capabilities. As described above, another strong need to develop small devices that work on site comes up. To solve this problem, *Reservoir Computing* is proposed. In short, normally *RNN* carries out a large number of nonlinear transformations. All links are variable nonlinear operator (neurons). They change their weights. (Fig. 8.13a). But in *Reservoir Computing*, at the early stage called "*Reservoir*", the weights of nonlinear operators are fixed and only at the last stage, linear transformation is carried out, i.e., details are adjusted at the last stage (Fig. 8.13b).

Thus, the part in the middle is regarded as "Blackbox". To describe it another way, normal RNN deals with each wave behavior, but *Reservoir Computing* does not care each wave behavior, but looks at the whole pond. Thus, it is called *Reservoir*. So, in *Reservoir Computing*, a large amount of reduction of computing becomes possible. Another benefit is rough evaluation becomes possible. For strategic management, what is more needed is a holistic evaluation, not a rigorously exact evaluation, i.e., rough sketch is more important for management than detailed one. Thus, *Reservoir Computing* shift computing from tactics focused to strategy focused.

This shift enables us to develop many physical sensors, such as micro photonics, and spin waves. Thus, *Reservoir Computing* is nothing other than *Natural Computing*.

Thus, performance evaluation is critical in *Reservoir Computing*. Mahalanobis Distance-Pattern (MDP) approach proposed here is expected to work effectively.

The above discussion is about the computing system. But from the standpoint of the social system, the world is changing. The Industrial Revolution introduced Division of Labor, so people started to work for others for external rewards. And

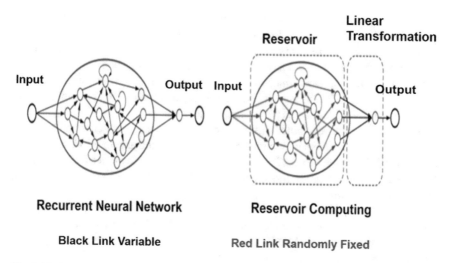

Fig. 8.13 Recurrent neural network (**a**) and reservoir computing (**b**)

as people wanted similar kinds of products, mass production prevailed. So, there is no benefit of living in the countryside, people are rushing into work in the city. The more people gathered, the more convenient the society became. So, the large number of people continued to flow into the cities.

But the industry framework changed from the centralized linear system to the parallel processing distributed system. Thus, yesterday, final product companies are dominating the market, but today intermediate components or parts companies are dominating. And the shift of our world to open world without boundaries reminded people of the benefits of living in the countryside. They can utilize the local resources and the cost performance of living is far less there in the countryside. You can explore to your heart's content, what you desire. So, people are spreading out more and more. Thus, the social system also shifts from centralized to distributed. But we should be careful. We are spreading out widely, but we are moving toward the *Connected Society*. Yesterday, we were connected because we lived in the same area, but today we are connected through remarkable progress of communication. Distance is not a problem anymore. Now, we are freed from area constraints,

8.7 Mahalanobis Distance-Pattern (MDP) Approach

Taguchi developed MD Pattern approach, but it is static pattern matching. *MDP Approach*, on the other hand, deals with dynamic issues of *Movement Coordination*. Interestingly enough, although they deal with the same issue of human movement, it is called *Motion Control* when we observe from outside, and *Motor Control*, if we pay attention to the inside of our body.

As Bernstein's cyclogram shows, our movement trajectory is fixed at the latter stage. So, it is straightforward to apply rational approaches to this stage. But at the early stage, our trajectories vary very widely. Therefore, research on "Coordination" is important. We coordinate many body parts and balance our body.

Even if we can observe trajectories only from outside, our movements can be captured as images. Then, we can evaluate our performance. MD combined with image patterns provides such a performance indicator.

MDP Approach is very much different from *MTS* because it focuses on dynamic changes of our movements with attention paid to speed and acceleration. And evaluation of the pattern is left to our *Instinct*. It should also be added that MDP accommodates both Motion and Motor movements, because MDP approach leaves perception and decision to Instinct.

MTS utilizes a typical pattern called *Unit Space*. And sample patterns are compared with Unit Space and if MD is smaller than the threshold value, then the sample pattern is identified as the same. If MD is larger than the threshold, then the sample pattern is different. So, the decision is automatically made.

But in the case of *MDP Approach*, the user evaluates his or her performance based on the performance indicator and recognizes what he or she should do to improve movements. How and where are left to the user's *Instinct*. The octopus interacts with the outside world without any prior knowledge but can deal with any environments and situations. They are using their *Instinct*. The stimuli of the outside world is the same for humans and the octopus. Humans depend too much on *Knowledge*, whereas the octopus makes the most of their *Instinct*.

If we think carefully, we, humans, depend on *Instinct* when it comes to *Strategic Decision-Making*, because strategy is nothing other than finding a goal. Just an accumulation of pieces of *Knowledge* does not work. We need to explore the new horizon to achieve the goal.

This is the difference between *Control* and *Coordination*. *Control* is explicit and rational. Thus, it is deeply associated with *Automation*. The whole scenario is built upon the idea of reproducibility. But *Coordination* is the efforts to make the most of the current resources. And judgment varies from person to person. If the intended goal is achieved, that's fine. It does not matter it may not be the best or the optimum. To borrow Simon's words, *Coordination* is the world of *Satisficing* [12]. And, to borrow Shakespeare's words, *Coordination* is "All's well that ends well" [13]. In short, *Coordination* is the world of *Pragmatism*. The World 2.0 pursues the World of *Pragmatism* with the final goal of making your life enjoyable.

There are many researches on *Motion control*, because it is observable from outside, but although its importance is very much emphasized, researches on *Motor Control* are still few and they study independently on different topics. So, how we coordinate and balance our body inside of us is still unclear.

But we combine *Motion* and *Motor* information together to move as we like. So, this is nothing other than *Strategic Decision-Making*. Thus, the shift from the World 1.0 to the World 2.0 may be described as an effort to make the most of our *Instinct* to make our life enjoyable. In fact, we are born with *Instinct* alone. We do not have *Knowledge*, when we are born. But babies explore the outside world and store

Knowledge. We, especially engineers, have been focusing too much on *Acquired Abilities.* We forgot *Inherent Abilities.* We should make the best use of our *Inherent Abilities.* Fukuda published "Self Engineering: Learning from Failures" [14]. The details of *MDP Approach* are described in this book.

This book asserts that you should stop working for others and regarding work as forced labor. You should start working for yourself to make your life enjoyable. That is the shift from the World 1.0 to the World 2.0. To realize such a shift, we need to develop the following two.

(1) Smart machines that adapt to our personal way of working in terms of decision-making, movements, actions, etc.
(2) Environments that motive us to work for ourselves.

8.7.1 Applications

MD Pattern (MDP) Approach works very well to meet such requirements mentioned above. It will facilitate developing *Wearable Robots* such as shown in Fig. 8.14.

Fig. 8.14 Wearable robots

Fig. 8.15 Wearable robots

> **Swimming**
> **Weather Adaptive Shirt**
> **Prosthetics**
> **Rehabilitation**
> **Healthcare**
> **Assist Robot**

Fig. 8.16 Physical arts

> **Dancing**
> **Figure Skating**
> **Sports**

Other applications are also shown in Fig. 8.15. As described many times in the above, *MDP Approach* can solve the problems whose parameters cannot be identified. Swimming is a typical example. Another application would be *Textile Temperature Sensors*. They enable us to develop a shirt which expands when it is hot, and which shrinks when it is cold. It may be called a very personal air conditioner. In artificial legs or arms field, the interaction between these artifacts and human bodies is very often discussed. But if we note that our body builds are different from person to person and how we move and feel are different from person to person, *MDP Approach* will serve a great deal to find the best break in conditions so that patients can move comfortably in their own way. This would bring them great happiness. Such careful attention toward individual characteristics can be well taken care of by *MDP Approach*. Therefore, it will bring satisfaction and happiness to patients in rehabilitation and in health care.

It should also be emphasized that *MDP Approach* will contribute to create physical arts. Dancing and Figure Skating are typical examples. But all Sports are in essence physical art. So *MDP Approach* will motivate us to creating physical arts to enjoy life (Fig. 8.16).

References

1. https://en.wikipedia.org/wiki/Prasanta_Chndra_Mahalanobis
2. https://en.wikipedia.org/wiki/Mahalanobis_distance
3. https://en.wikipedia.org/wiki/Mycin
4. M. Lukosevicius, H. Jaeger, Reservoir computing approaches to recurrent neural network training. Comput. Sci. Rev. **3**(3), 127–149 (2009)

5. G. Tanaka, T. Yamane, B.J. Heroux, R. Nakane, N. Kanazawa, S. Takeda, H. Numata, D. Nakano, A. Hirose, Recent advances in physical reservoir computing: a review. Neural Netw. **115**, 100–123 (2018)
6. V. Kostov, S. Fukuda, M. Johansson, Method for simple paralinguistic feature in human face. Image Vis Comput. J. Inst. Image Electron. Eng. Jpn. **30**(2), 111–125 (2001)
7. V. Kostov, S. Fukuda, Emotional coding of motion using electromagnetic 3D tracking instrument. Appl. Electromag. Mater. Sci Device Jpn. J. Appl. Electromag. Mech. **8**, 229–235 (2001)
8. G. Taguchi, S. Chowdhury, Y. Wu, *The Mahalanobis-Taguchi System* (NY, McGraw-Hill Professional, New York, 2000)
9. G. Taguchi, R. Jugulum, *The Mahalanobis-Taguchi Strategy: A Pattern Technology System* (Hoboken, NJ, Wiley, 2002)
10. https://en.wikipedia.org/wiki/Facial_Action_Coding_System
11. https://en.wikipedia.org/wiki/Recurrent_neural_network
12. https://en.wikipedia.org/wiki/Satisficing
13. https://en.wikipedia.org/wiki/Alls_,Well_That_Ends_Well
14. S. Fukuda, *Self Engineering: Learning from Failures* (Springer, London, 2019)

Chapter 9
Think Big, Make It Simple: Enjoy the Process

The Industrial Revolution introduced division of labor and we started to work for others, i.e., we started to work for external rewards. Efficiency and cost performance were emphasized. So, it soon led to mass production and customers were called consumers. In fact, Adam Smith said "consumption is the sole end and purpose of all production". Thus, we were driven to mass production. And products were evaluated based on social status, pride, etc.

The Industrial Revolution accelerated technology progress and engineers pursued higher and higher technologies. As we enter the twentieth century, technologies were accelerated to advance rapidly in an exponential way. Therefore, it led us to the technically *Divided Society*. In each technological sector, higher and higher were the keywords, but they were independent of each other and there were few efforts to integrate them.

For example, in the transportation industry, airplanes, ships, trains, automobiles, etc. were developed and they made remarkable progress in each industrial sector. But when we said "I would like to fly like a bird", we wished, of course, to fly like a bird, but what we really dreamed was to fly, walk and swim like a bird and to travel comfortably across any environments. The purpose of engineering is to make our dreams come true. But our dreams were completely forgotten, and advancement of technologies has been pursued with tremendous efforts.

Thus, our society has been producer-centric. Customers have been always called *Consumers*. They have been considered only as labor providers and product consumers. No attention has been given to what will provide customers with true pleasure. The word *Culture* comes from *Cultivate*. We create *Culture*. The word *Consumer Culture* is contradictory. We forgot what customers really want and what provides them with joy.

The word "customer" literally means what they really want. Each person would like to *customize* his or her lifestyle. Everybody would like to personalize his or her life. But engineering up to now has been paying efforts to make everybody's life the same. Thomas Friedman's book "The World is Flat" [1] expresses such an idea.

© The Author(s), under exclusive license to Springer Nature Switzerland AG 2020
S. Fukuda, *World 2.0*, SpringerBriefs in Applied Sciences and Technology,
https://doi.org/10.1007/978-3-030-51588-1_9

But we should remember *Diversification and Personalization* are rapidly increasing importance.

As described earlier in 7.1 Self is our Mainspring, *Personalization* is our intrinsic desire. We are born to develop *Self*.

The current situation is we are now thrown into a river. Water is changing every minute. We cannot identify parameters, so we cannot control our movements, but we need to swim against the flow to reach our goal. *Adaptability* is emphasized these days, but what is more important is to outsmart the flow and win the game to achieve our goal.

Sustainability is emphasized these days. But what is important is *Survival*. We cannot stay where we are. We are born to grow, and we need to survive the frequently and extensively changing outside world. We should live for tomorrow. We should stop living for now.

9.1 From Product to Process: From Knowledge to Wisdom

This boils down to the fact that now processes become more important than products. In an age when product value was highly evaluated, how we can realize a flat society was important. But today, processes become more important. And it becomes crucially important that we outfit the environments and win the game. How we do it varies widely from person to person. In other words, *Wisdom* becomes more important than *Knowledge*. We need to pay more attention to our *Instincts*. We should learn from the octopus.

The World 2.0 pursues "How we can live naturally and enjoy life". And we need to team up with machines to make our life pleasurable and enjoyable. As IoT holds, we need to remove the wall between living and nonliving things, and we must develop emotional chemistry and work together on the same team. But we must not forget that it is us, humans, who make decisions.

9.2 From Innovation to Invention

Rolf Faste, Stanford Professor, told us that *Innovation* is to expand the yolk or the yellow of an egg. But *Invention* is to create another yolk on the same egg. *Invention* is closely associated with *Situational Awareness* or *Situational Consciousness*. *Innovation* is fundamentally extension of the present world, but *Invention* is closely related to outsmarting or how we can win the game. As described in Sect. 5.3 Soccer as an Example, soccer yesterday pursued *Innovation*, but soccer today needs *Invention* all the time.

In the World 1.0, people were working for others and what was expected from them was to innovate. It was a world of products. But in the World 2.0, the main emphasis is to create joy and satisfaction. Let us invent to enjoy.

Well, restaurant dishes taste very good. But these dishes call for a good chef, good tools and good materials. But suppose we take restaurant dishes every day. Do you feel happy and satisfied? I would say, "No". Instead, think of the leftovers in your refrigerator. If you cook them in a new way or combine them in a different way and if this dish tastes very good, can you imagine how much larger satisfaction and happiness you will get? This is because this dish comes from internal motivation and self-decision. As *Self-Determination Theory (SDT)* points out, we humans feel the maximum satisfaction and happiness if we do things out of our *Self*. No external rewards can provide them.

In the World 1.0, we were running on the same track, making efforts to go further and further. But in the World 2.0, we should look for other tracks. We should make efforts to go wider and wider. It is the world of *Exploration*. We should pursue to shift from high technology to low technology. We should make efforts to find how we can make the most of the current resources. Creating a new dish using leftovers is one example. Electric Vehicles (EV) is another example. The parts of EV are interchangeable. So, we can enjoy the same way as we do with accessories or toys. If we can make the processes easy enough so that everybody can enjoy the process of producing a product, then we can create a world filled with psychological satisfaction and happiness.

This is the joy combination brings. Let us consider another example. 3D Printing. There are many discussions about it. But most of it is how we can use the technology to achieve more sophisticated products. These people are interested in going further on the same track. But if we look at this technology from a customers' perspective, it will provide much greater satisfaction and pleasure, if they can make what they want by themselves. In fact, *Makers* [2] demonstrates how such attempts are rewarding. Engineers should pay more efforts to make the current technology simpler enough so that non-experts can enjoy the process. In short, "*Self-Realization as a Service (SRaaS)*" will be the business keyword in the next World 2.0.

Another business keyword in the World 2.0 may be "*Customization as a Service (CaaS)*". Engineers are expected to produce such smart machines which have warm consideration to personal behaviors and characteristics of humans. Then, we can enjoy working together with machines on the same team, because machines perceive our personal preferences or differences and make their best effort to perform accordingly.

References

1. T. Friedman, *The World is Flat: A Brief History of the Twenty-first Century* (NY, Farrar, Straus and Giroux, New York, 2005)
2. https://en.wikipedia.org/wiki/Maker_culture

Chapter 10
Motivation: From Material to Mental

As described in detail in Chap. 1. EVOLVING WORLD, the culture in the World 1.0 is *Consumer Culture* and people pursued satisfaction with products. But as Maslow pointed out, human needs gradually shift from material satisfaction to mental satisfaction. So, people are now looking for mental satisfaction. In short, people's value used to be homogeneous and it was the value of products, but now it becomes widely diversified and personalized and the value of processes is rapidly increasing.

In the World 1.0, people shared the same criterion of value. The value was directly linked to social status. But in the World 2.0, people come to have their own criterion due to rapid progress of diversification and personalization. So, what interests them is not social status, but how much they get the sense of accomplishments from working for themselves. It is very much of a personal value, which varies from person to person.

Charles Darwin proposed "Survival of the Fittest" [1]. We often misinterpret this word that the person who is most excellent will survive. This is a complete misunderstanding. No, sometimes those who are left behind in the usual race win the game of survival. Nature accepts those who fit or adapt best to the outside world. And nature changes unpredictably. So, who knows who will really fit the changing environments and situations? It is up to nature who to select. It is *Natural Selection* [2].

Why we, humans, want to actualize ourselves at the top of Maslow's hierarchy of human needs is because human progeny might survive if Nature selects your *Self* to survive. We must pay attention to the difference between *Selfish* and *Self-Actualization*. *Selfish* is just doing things for yourself alone. You don't care about progeny. If you feel good, then, you are happy. Everything is just for yourself. You live for today. You don't care tomorrow. You live for your own life alone.

Self-Actualization may look the same, but in fact it is very much different. We are trying to demonstrate how capable we are as an individual, but personality, body builds, etc. vary widely from person to person. This way by diversification, we are increasing the possibility of survival of the human species. You don't think you are working for human species. But working for yourself this way is contributing to the

S. Fukuda, *World 2.0*, SpringerBriefs in Applied Sciences and Technology,
https://doi.org/10.1007/978-3-030-51588-1_10

survival of the human species. Of course, nature might select someone else. But from the standpoint of the whole human species, you are contributing to the survival and expansion of the human species.

Deci and Bryan proposed *Self Determination Theory* and made it clear that we get the greatest satisfaction and happiness, when we do the job out of our internal motivation and by self-decision. No external rewards can provide such amount of satisfaction and happiness.

We must remember that *Motivation* and *Emotion* come from the same Latin "movere". So, *Emotion* means e = ex-motion, i.e., move out. When we are motivated, we move out into the real world to explore. So we live for tomorrow.

Evolution is a matter of human species. *Growth* is, on the other hand, a matter of yourself. Deci and Ryan pointed out also the importance of our needs for growth. Working for yourself satisfies both needs as a human and as an individual.

References

1. https://en.wikipedia.org/wiki/Survival_of_the_fittest
2. https://plato.stanford.edu/entries/natural-selection

Chapter 11
Live for Tomorrow: A Rising Tide Lifts All Ships

Living things are born to grow. But growth is a personal matter. As a species, living things evolve. They adapt to the changes of the outside world to survive, and they expand the world of their species.

Why are living things so different from each other in the same species? It is because the more they are diversified and personalized, the more the possibility of survival and expansion of the world of their species increases.

The World 1.0 pursued homogeneity and reproducibility and we worked for others for external rewards. But the World 2.0 pursues *Self*. We "work" to realize *Self*. It brings us psychological pleasure and happiness which no external reward can provide. Such personal pursuit of pleasure and happiness may be compared to waves. There are many different waves at sea. But there are tides. We should remember the word, "A rising tide lifts all ships". What the World 2.0 pursue is not only making waves only personally, but making tides rising so that all ships (all of us) will be lifted, i.e., individuals and humans as a whole grow and evolve together.

In the World 1.0, area was the basis of all things. But in the World 2.0, we will be living all across the Earth. Our world is changing from a small, closed world with boundaries to wide, open world without boundaries. Therefore, we will be distributed on earth (or maybe in space) and "work", wherever we like and whatever we like to do. We can utilize the local resources there.

We should remember that with the progress of Edge Computing and IoT, small devices (sensors/actuators) become available and with the progress of *Virtual Reality* (Note this Virtual means Value), we can share the same experience, no matter where we are. Smell, haptics, etc. of the outside world can be shared soon, no matter how much far away we may be.

4C's is a famous keyword, but its definition is different from sector to sector. So, here is 4C's for the World 2.0.

(1) Community
(2) Communication
(3) Collaboration
(4) Creativity.

© The Author(s), under exclusive license to Springer Nature Switzerland AG 2020
S. Fukuda, *World 2.0*, SpringerBriefs in Applied Sciences and Technology,
https://doi.org/10.1007/978-3-030-51588-1_11

Community is now a global community. It used to be a local community in the World 1.0, but in the World 2.0, we can develop community without any constraints. The importance of communication can be easily recognized, if we remember IoT. IoT means Internet of Things and Internet means Communication. Communication is important to link nodes to form adaptive network. In fact, links play the role of messenger. And needless to say, the World 2.0 will be the world of adaptive network. So, collaboration is indispensable. Well, someone might ask why creativity is here. It is because we will create the new World, World 2.0.

The World 1.0 was the world of *Innovation*, but the World 2.0 will be the world of *Invention*.

Let us create the new world and enjoy working. Challenge is the core and mainspring of all human activities. It is a big challenge. Create work for you and enjoy.

Printed in the United States
By Bookmasters